CRIME
AND
EVERYDAY LIFE
FOURTH EDITION

CRIME
AND
EVERYDAY LIFE

FOURTH EDITION

MARCUS FELSON | RACHEL BOBA

Rutgers University   *Florida Atlantic University*

Los Angeles | London | New Delhi
Singapore | Washington DC

For information:

SAGE Publications, Inc.
2455 Teller Road
Thousand Oaks, California 91320
E-mail: order@sagepub.com

SAGE Publications Ltd.
1 Oliver's Yard
55 City Road
London EC1Y 1SP
United Kingdom

SAGE Publications India Pvt. Ltd.
B 1/I 1 Mohan Cooperative
 Industrial Area
Mathura Road, New Delhi 110 044
India

SAGE Publications
 Asia-Pacific Pte. Ltd.
33 Pekin Street #02-01
Far East Square
Singapore 048763

Printed in the United States of America

Library of Congress Cataloging-in-Publication Data

Felson, Marcus, 1947-
Crime and everyday life/Marcus Felson, Rachel L. Boba. — 4th ed.
   p. cm.
Includes bibliographical references and index.
ISBN 978-1-4129-3633-0 (pbk.: acid-free paper)
   1. Crime—United States. 2. Crime prevention—United States. 3. Social control.
I. Boba, Rachel. II. Title.

HV6789.F45 2010
364.973—dc22   2009033512

This book is printed on acid-free paper.

09   10   11   12   13   10   9   8   7   6   5   4   3   2   1

Acquisitions Editor:	Jerry Westby
Associate Editor:	Eve Oettinger
Editorial Assistant:	Nichole O'Grady
Production Editor:	Catherine M. Chilton
Typesetter:	C&M Digitals (P) Ltd.
Proofreader:	Sue Irwin
Indexer:	Molly Hall
Cover Designer:	Edgar Abarca
Marketing Manager:	Jennifer Reed Banando

BRIEF CONTENTS

———•◦•———

except that, unfortunately, someone happened to die. Murder has two central features: a gun too near and a hospital too far. Tedeschi and Felson (1994) find many parallels between murder and simple fights. Although some murderers intend to kill from the outset (see Felson & Messner, 1996), even they usually have simple reasons not worth televising. More commonly, the offender intended to harm the victim, but did not know how much.

The Mass of Minor Offenses

Yet the vast majority of times, the harm is limited. The picture we are painting is well supported by the National Crime Victim Survey (Bureau of Justice Statistics [BJS], 2009). Property crime victimizations far exceed violent victimizations, with the simplest thefts and burglaries most common. Confirming the same general conclusion, self-report surveys pick up a lot of illegal consumption and minor offenses, but little major crime. For example, high school students admit considerable underage drinking, minor theft, and plenty of marijuana experimentation, but only a small percentage report using cocaine or hard drugs. The Monitoring the Future Survey for the year 2006, for instance, asked high school seniors about their use of various drugs. Some 31% reported having tried marijuana or hashish that year, but only 4% reported using Ecstasy. When students are asked about use in the past 30 days, these numbers fall to much lower levels (Johnston, O'Malley, Bachman, & Schulenberg, 2007). This principle holds strong: Minor drugs far exceed major drugs, and occasional usage far exceeds regular usage. None of these drugs ever does nearly as much harm, in percentage terms, as simple alcohol.

You can see, then, that most offenses are not dramatic. Property crime far exceeds violent crime. Violent crimes that do occur are relatively infrequent and leave no long-term physical harm. When injury does result, it is usually self-containing and not classified as aggravated assault, much less homicide. Real crime is usually not much of a story: Someone drinks too much and gets in a fight. There is no inner conflict, thrilling car chase, or life-and-death struggle. He saw, he took, and he left. He won't give it back.

This is not to deny that dramatic events occur in real life. By the time this book is in your hands, another teenager may shoot up a school, and people will then talk about this one event as if it represented all crime. Keep your focus on the plain facts of crime and ignore the dramatic event of the month.

2. THE COPS-AND-COURTS FALLACY

It is easy to exaggerate the importance of the police, courts, and prisons as the key actors in crime production or prevention. But we have to keep reminding people that crime comes first, and the justice system follows. The *cops-and-courts fallacy* warns us against overrating the power of criminal justice agencies.

Police Work

Real police work is by and large mundane. Ordinary police activity includes driving around a lot, asking people to quiet down, hearing complaints about barking dogs, filling out paperwork, meeting with other police officers, and waiting to be called up in court. If you ever become a crime analyst, you will quickly learn that

- Many calls for service never lead to a real crime report.
- Many complaints (e.g., barking dogs) bother a few citizens, but do not directly threaten the whole community.
- Many problems are resolved informally, as they should be.

To quote the standard line, police work consists of hour upon hour of boredom, interrupted by moments of sheer terror. Some police officers have to wait years for these moments. Most seldom—or never—take a gun out of its holster. Most are never shot at and never shoot at anybody else. In 2007, police shot and killed 388 citizens in the line of duty (FBI, 2008a). Some 57 officers were feloniously killed in the line of duty, and another 83 died in accidents (FBI, 2008a). These are large numbers compared to European nations, but small numbers in terms of overall police activity and number of officers—just under 700,000 in 2007 (FBI, 2008b). Of course, a police officer has a right to be upset about every threat to any officer. Overall, however, their mortality rates are not high. Their more common problems are rude encounters with people who cannot keep their mouths shut.

The short arm of the law is best explained by one fact already noted: Most crimes are never reported to the police in the first place. In the 2005 National Crime Victim Survey, about 57% of those who said they were victimized also said they had not reported the crime to the police (BJS, 2009). Self-report studies cited earlier in this chapter turn up even greater numbers of illegal acts

Very active offenders tend to continue offending as they get older (see Nagin, Farrington, & Moffitt, 1995). These offenders, who tend to engage in extended substance abuse, are described in Curve B of Exhibit 1.1. Although this group is small in number, it carries out far more than its share of offenses. Could these long-active criminals be the ones corrupting the young? That might sometimes be the case. But Curve B shows that this long-active group also initiates criminal behavior at even younger ages, getting even deeper into trouble during adolescence. The main corrupters of youths are other youths.

We have heard many people declare that "Prisons should keep young offenders separate, or else the hard-bitten criminals will be a bad influence on them." But young inmates are the ones causing the most trouble inside the prisons, which separate by age to protect older prisoners from young thugs. Juvenile homes are also settings for very serious internal offending, such as frequent assaults by those in their early or mid-teens. Indeed, juvenile homes must keep constant watch to keep the stronger boys aged 12 to 14 from raping the weaker boys.

Some people believe that the real problem is incarceration itself, whether in juvenile or adult facilities. One finds excellent *ethical* arguments against excessive or premature punishment (see von Hirsch, 1987). But that does not prove innocence of youth. To understand the impact of any incarceration in practical terms, consider the standard lockup sequence:

1. A youth lives a risky lifestyle among dangerous people,

2. is incarcerated for a certain time, and

3. returns to the same risky lifestyle and dangerous people.

To people who hang out in libraries, the middle stage seems so hellish that it could only have ruined a person stuck there. But a dangerous life before and after is hellish, too—maybe more so. Again, it is difficult to defend the hypothesis that a largely innocent youth was ruined by the system.

Some of the most severe offenders may look old but are really rather young. A dangerous lifestyle, including prolonged substance abuse, takes its toll on the body. Those aged 18 sometimes look 22; those 22 might look 28; those 28 look 38; those 38 look 58. That helps to reinforce the misconception that bad guys are relatively old. On the other hand, substance abuse can keep people thin, sometimes masking bodily deterioration for a while.

5. THE INGENUITY FALLACY

The false image of "the criminal" derived from the media also creates an *ingenuity fallacy*. To be good foils for the hero, criminals must be almost as crafty and tough. Consider Professor Moriarty, the evil and ingenious criminal in an epic struggle with Sherlock Holmes. It would hardly be fair to send the brilliant Holmes chasing after a drunken fool. Our ideas about classic criminals also include the skilled cat burglar who can slip into a third-story room of sleeping victims, quietly pocket valuables, then glide down the drainpipe with nary a worry of excess gravity. Edwin Sutherland (1933) chronicled his interviews with Chick Conwell in *The Professional Thief, by a Professional Thief.* This offender knew how to switch fake jewels for real ones right in front of the jeweler, and had the skill to trick people out of their money. Pickpocketing and safecracking were not problems for him.

However, most crime as we know it today needs no advanced skills. Maurice Cusson (1993) of the University of Montreal writes about the "de-skilling" of crime in recent decades. Technology has made modern safes too difficult to crack, and most people avoid having that much cash anyway. Crowds suitable for old-fashioned pocket picking decline in a suburban society. Dermott Walsh (1994) has generalized this issue in what he calls the "obsolescence of crime forms." He lists 24 offenses that have become largely obsolete, including bribery of voters, eavesdropping, and illegal abortion. The other side of the coin is that modern life has produced many easy crimes (Cohen & Felson, 1979; Felson & Cohen, 1980). Lightweight durables are easy to steal, and more cars on the street have been simple to break into. Houses empty during the day with valuable items easy to steal make it hardly worth the trouble to crack a safe. Technology, such as the Internet and online shopping, make theft of credit card numbers easier and more lucrative. Chapter 3 discusses offender decisions further, and Chapter 10 explains how preventive changes might add to the list of obsolete crimes.

Part of the ingenuity fallacy stems from the embarrassment of the victim. It is difficult to admit how foolish you were in leaving yourself open to the offender. That's why you might be tempted to tell yourself, police, and anyone who asks, "A *professional* criminal broke into my house." People who hide the jewels in the cookie jar or the money in the bathroom think that nobody else ever thought of those hiding places. Remember that the burglar's mother also hides the jewels in the cookie jar. If you were looking for someone else's valuables, where would you look?

One proof that most crimes are not ingenious is the tiny slice of time they require. Many a theft is carried out in ten seconds or less. A burglary takes a minute or two, sometimes as much as five, but sometimes an unlocked door helps the burglar get in and out in half a minute. An open garage takes almost no time to slip in, take a bicycle, and ride away. If the burglar enters a building, he still acts quickly to avoid detection, perhaps making a mess rummaging through the victim's belongings in search of loot. Robbery also is a quick crime. If someone points a gun at you and asks, "Your money or your life?" how long does it take you to decide? An offender can easily figure out how to get others to comply with his wishes. That's why most crimes involve so little planning, plotting, or creativity (see Chapter 3).

6. THE ORGANIZED-CRIME FALLACY

Organization, drama, and crime families—these are the images of organized crime and groups of criminals. These images might have applied in Sicily, but criminal conspiracy seldom works that way in North America today. The *organized-crime fallacy* is the tendency to attribute much greater organization to crime conspiracies than they usually have.

Crime Conspiracy

The three basic principles of crime conspiracy are

1. Act quickly to escape detection and minimize danger from other offenders.

2. Have direct contact with as few co-offenders as possible to avoid betrayal.

3. Work as little as possible to get a lot of money.

Given these principles, large groups and organizations make no sense at all for most types of crime. This is not to deny that *some* organized crime exists, and that some primordial organization eventually reaches more advanced levels of organization (see Felson, 2009). But that is not the normal process, even for criminal conspiracies.

Instead, most criminal conspiracies work like a chain letter. Perhaps Joe grows marijuana and sells some to Mary. She distributes smaller packages among five others. They break packages down more for a few more people, and so on (see Felson, 1998, Figure 3.7). Each of these packages is handed off very quickly. This illegal network may involve many people, but few of them know each other. If one is arrested, only one or two others might be incriminated. Peter Reuter (1998), an economic expert in illicit markets, shows that drugs and gambling have much simpler organizations than their popular image. Don't forget how easy and quick it is to hand someone a package in return for money, to take a bet, or to sell quick sex. There's no point being a criminal if you have to go to long meetings.

Juvenile Street Gangs

Juvenile gangs have a remarkable image as cohesive, ruthless, organized groups of alienated youths who dominate local crime, do the nation's drug trafficking, provide a surrogate family, and kill anybody who quits, which has led the public to misunderstand the more common dangers.

The leading expert on juvenile gangs, Professor Malcolm W. Klein, started by studying gangs face to face (Klein, 1971). He expected to find coherent groups of boys involved in exciting things. Instead, he found groups with very loose structures: people fading in and out, and frequent disintegration. Most of the time, they were extremely boring. Klein described the street gang as an onion, with each part peeling off to reveal another part, then another, until you got to the core. The few core members were more active than the others (i.e., they hung out regularly, doing next to nothing). Yet most members were peripheral—there one day and not the next. Surprisingly, Klein's experience revealed that the very social workers who were supposed to help boys get away from gangs were actually keeping the gangs cohesive. Gangs with no social workers to help them fell apart even more often. Klein's (1995) landmark follow-up book, *The American Street Gang: Its Nature, Prevalence, and Control,* punctures many preconceptions about gangs (see also Klein, Maxson, & Cunningham, 1991). Klein and his associates were not persuaded by the shows put on by the gangs, media, or police press releases.

For several reasons, juvenile street gangs are not all they are cracked up to be (Felson, 2006). First, the word *gang* is overused to describe groups of high school kids, motorcyclists, skinheads, drug dealers, and so on. Second, gangs have nongang features—individuals have jobs, go to school, are friendly—and spend more of their time hanging out than committing crime. Third, gangs are often credited with crime committed by other youths, and their numbers are often inflated. To be sure, juveniles associated with gangs are involved in substantial amounts of crime (Rosenfeld, Bray, & Egley, 1999), and crime may be part of their identity (Tittle & Paternoster, 2000), but most of their crime is petty and very local. In his police problem-solving guide on young offender gun violence, Anthony Braga (2003) concludes, "Even in neighborhoods suffering from high rates of youth gun violence, most youth are not in gangs" (p. 5). He cites researchers in Boston and Minneapolis who have found that gang members represented less than 1% of all youths and less than 3.5% of residents between the ages of 14 and 24, respectively (Braga, 2003). Fourth, different gangs use the same name to create the perception of a larger organization— Bloods and Crips of Los Angeles, California, and Omaha, Nebraska. Finally, gangs are unstable, and often do not survive as a continuous group over time. Just as other teenagers join and leave cliques, so do juvenile gang members.

So, let us describe a typical street gang, the Undependables. Membership is volatile, even within its core. It is loosely linked to a neighborhood and has even looser links to nearby gangs, also called the Undependables. There may be the Bay Street Undependables and the Madison Park Undependables, who seldom talk and may even fight each other. There may also be Wannabes, boys who imitate the gang or would like to join it.

The Undependables have an extra presence in local crime, but a majority of such crime is done by youths who are not "members." When a small group of Undependables steals something, they do not share it with the group but keep it for themselves. Nor can they be depended on to keep crime within the group. For example, two members may well get together with two not in the gang to break into a house. Although some members may play a role in selling illegal drugs, the gang as a whole is not organized to sell drugs,[2] despite its reputation for doing that. A juvenile gang may do evil, but it seldom does *cohesive* evil. Indeed, one of the fascinating features of crime is that so much harm can be done with so little togetherness.

Maybe it's too hard to accept the idea that a few youths can do so much harm. It is easier to attribute their actions to evil gangs or to organized crime, drawing juveniles into their grandiose purposes.

7. THE AGENDA FALLACY

The *agenda fallacy* refers to the fact that many people have an agenda and hope you will assist them. They want you to take advice, vote a certain way, or join their religious group. They may be totally sincere, but still they have plans for you. Their promise, usually bogus, is that their agenda will greatly reduce crime in society.

Moral Agendas

Many people believe that declining morality is the cause of crime. Hunt (1999) uses the phrase "the purity wars," and other criminologists have referred to "moral panics" as interpretations of crime and other social problems. The basic moral sequence is supposed to be as follows:

1. Teach and preach morality to people.

2. They then do what's right in practice.

3. That prevents crime.

If this sequence were correct, more teaching and preaching would prevent crime. A moral agenda would then be justified. Almost everyone has been taught morality, but many still commit crime. Hypocrisy is too powerful in human nature. People can easily talk about good without doing it. They can even believe in good without putting their pristine beliefs into clean action (Hirschi & Stark, 1969). As Dennis Wrong (1961) explained in a classic article, norms (moral standards) do not guarantee moral behavior. Nor does immoral behavior prove a lack of moral training. For example, the high murder rate in the United States does not prove that Americans believe in murder or that they are trained to commit murder. If that were the case, why do U.S. laws set such high levels of punishment for murder? Why would U.S. public opinion show such outrage at murderers and other serious criminals? And why would U.S. nonlethal violence rates be lower than those of so many industrial countries?

Consider a parallel question: Why do people become overweight? They don't want to be fat. They aren't trained to be fat. They don't need to be preached at. It would help if they did not live in a nation where food is rich and its prices cheap, and daily work burns few calories. The point is simple: People don't always follow their own rules.

This is not arguing against trying to instill morality. It's good to teach right from wrong, but you cannot really expect other people to do what you tell them when you aren't watching. On the other hand, morals do play a role in society. Each of us knows the rules and that someone might turn us in for breaking them. Morals give Joe a license to watch Peter, and Peter a license to watch Joe.

Religious Agendas

Many religious groups feel that conversion to their faith or values will prevent crime and that failure to follow will lead to more crime. Yet some of the most religious regions of the United States have very high crime rates, and the greater U.S. religious observance (compared to Europe) has not given us lower homicide rates. Hartshorne and May (1928–1930) found that young people in religious schools were just as likely to lie and cheat as those in public schools.

Yet *some* studies find correlations between religious activity and avoidance of crime. We think these correlations can be explained by the demands of sitting quietly in a church or synagogue, or listening to a sermon. Those most inclined to break laws have trouble sitting, so they stop going to church or never even start. Later, the researcher comes along and finds this correlation between churchgoing and offending, concluding, erroneously, that religion must have reduced crime among those who still go to church. Churchgoing and crime avoidance correlate for an entirely nonreligious reason: the presence of greater self-control (see Gottfredson & Hirschi, 1990).

But let's give religious institutions credit. They often do a better job of *supervising* people. A close watch on the flock keeps it from straying. Church schools tend to be smaller than public high schools, giving them more effective supervision of youths. They also kick out anybody who behaves too poorly. Smaller church groups can keep close tabs on their flock and thereby remove crime opportunities. Religious groups with quite incompatible beliefs might get somewhere in crime prevention by supervising young people closely. But they have the same problem as everybody else: Turn their heads and their young flock strays; and even the older flock needs some supervision.

Social and Political Agendas

A wide array of political and social agendas has been linked to crime prevention. If you are concerned about sexual morality, tell people that sexual misbehavior leads to crime. If you are a feminist, proclaim that rape is produced by antifeminism. If you dislike pornography, link it to sexual or other crimes. If the entertainment media offend your sensibilities, blame them for crime and demand censorship as a crime prevention method. If you are in favor of a minimum wage as part of your agenda, then why not argue that it will prevent crime? Right-wing, left-wing, or whatever your agenda, if there is something you oppose, blame that for crime; if there is something you favor, link that to crime prevention. If there is some group you despise, blame them and protect others; this is what Richard and Steven Felson (1993; also R. Felson, 2001) call "blame analysis." Joel Best (1999) goes so far as to write about "the victim industry," publicizing its sufferings in order to make claims on society. These are political tactics, not the way to study crime. Many crime reduction claims are far-fetched, even if the proposals are sometimes good.

Welfare-State Agenda

It's usually a mistake to assume that crime is part of a larger set of social evils, such as unemployment, poverty, social injustice, or human suffering. Consider the welfare state and crime. Some people hate the welfare state and blame it for crime. Others like the welfare state, promising that more social programs will reduce crime. We maintain that crime variations in industrial nations have nothing to do with the welfare state.

It is interesting to see partisans on this issue pick out their favorite indicators, samples of nations, and periods of history in trying to substantiate their assertions that rising poverty or inequality produce crime. Yet most crime rates went *down* during the Great Depression. We see all the economic indicators rising *with* crime from 1963 to 1975 (see Cohen & Felson, 1979; Felson & Cohen, 1981). We see the same indicators changing *inversely* to crime in the past few years in the United States.

Evidence of the mistaken welfare-crime linkage is evident by looking at the crime rate changes since World War II. Improved welfare and economic changes, especially for the 1960s and 1970s, correlated with more crime! Also, Sweden's crime rates increased 5-fold and robberies 20-fold during the very years (1950 to 1980) when its Social Democratic government was implementing more and more programs to enhance equality and protect the poor (see Dolmen, 1990; Smith, 1995; Wikstrom, 1985). Other "welfare states" in Europe (such as the Netherlands) experienced at least as vast increases in crime as the United States, whose poverty is more evident and whose social welfare policies are stingier. Clearly, something was happening in all industrial societies leading to a wave of crime that only recently has leveled off or been reversed.

America's welfare stinginess relative to Europe is often used to explain allegedly higher levels of crime and violence in the United States. It is hard to make international comparisons, when laws and police collection methods differ. But we now have a way to solve the problem. In a major scientific coup, Patricia Mayhew of the British Home Office and Professor Jan Van Dijk of the University of Leiden in the Netherlands negotiated throughout the world to get a single crime victimization survey translated and administered in many different nations. Results of that work are now showing that *the United States does not have higher crime victimization rates than other developed countries.* Nor is violence higher in the United States! In fact, the 2005 victim surveys in 15 industrialized countries (van Dijk, van Kesteren, & Smit, 2008) found this rank order in overall victimization:

1. Ireland
2. England and Wales
3. New Zealand
4. Iceland
5. Northern Ireland
6. Estonia
7. Netherlands
8. Denmark
9. Mexico
10. Switzerland
11. Belgium
12. United States
13. Canada
14. Australia
15. Sweden

Note in these cross-sectional data that the more generous welfare states of Europe, especially in Northern Europe, often have higher victimization rates than the United States. Additional data (van Dijk et al., 2008) on robbery, sexual assault, and assault with force show the United States having relatively modest rates of violence. The view that the United States is the violence capital is further undermined by research showing that school bullying is virtually a universal problem among nations (Smith et al., 1999). British police may carry no guns, but big British kids still beat up little ones (Phillips, 1991; Pitts & Smith, 1995; Smith & Sharp, 1994; Tatum, 1993).

Nonetheless, the United States has much higher *homicide* rates than any developed country of the world. How can we be moderate in general violence but very high in lethal violence? The presence of guns in the United States makes the difference (e.g., see Sloan et al., 1988; Zimring, 2001; Zimring & Hawkins, 1999). Americans are not more violent than Europeans; we just do a better job of finishing people off.

This is not an argument against fighting poverty or unemployment. Rather, it is an attempt to detach criminology from a knee-jerk link to other social problems. Crime seems to march to its own drummer, largely ignoring social injustice, inequality, government social policy, welfare systems, poverty, unemployment, and the like. To the extent that crime rates respond at all to these phenomena, they may actually increase with prosperity because there is more to steal. In any case, crime does not simply flow from other ills. As Shakespeare writes,

The web of our life is of a mingled yarn,
good and ill together.

—All's Well That Ends Well, Act IV, Scene 3

Crime has become a moral, religious, and political football to be kicked around by people with agendas. If you want to learn about crime, you do not have to give up your commitments, but keep them in their proper place. Learn everything you can about crime for learning's sake, not for such ulterior motives as gaining moral leadership, political power, or religious converts. If your political and religious ideas are worthwhile, they should stand on their own merits.

8. THE VAGUE-BOUNDARY FALLACY

Some criminologists believe that crime has no universal definition. They see crime as subjective, with society and its justice system "manufacturing" crime by changing the definition. Their intellectual lawlessness makes a mess of our field by

- Giving it no boundaries and keeping it vague
- Requiring a different criminology for each legal system
- Letting criminology students get an easy A, no matter what they write

The *vague-boundary fallacy* refers to the tendency to make criminology too subjective. It allows students and instructors to wriggle out of responsibility, and keeps crime science from developing.

A Clear Definition of Crime

Despite variations in laws and their application, we can define crime clearly and coherently for broad historical purposes:

A crime is an	*identifiable behavior*
that an	*appreciable number of governments*
has	*specifically prohibited*
and	*formally punished* (Felson, 2006, p. 35).

Thus, a criminal behavior defined *in broad historical terms* is not necessarily a *statutory* crime in all nations or all eras. This allows you to say, "Licensed prostitution is not criminalized in my country today, but is still on the historical list of crimes that applies to us all." Biologists will understand this well, agreeing that hedgehogs are animals, even though many countries don't have them.

For practical purposes, we may place any behavior on the crime list if it has been banned by at least 10 societies at any time in history, and at least 50 persons have been punished for that behavior in each of those societies. Although we might not agree about what behaviors *should* be criminalized, we can still agree that any given behavior has been treated criminally in enough places to be defined as a crime. On the other hand, behaviors that are seldom banned or almost never enforced are not included as crimes, even if they are written on the books somewhere. Thus, we can include variations among nations and historical periods and still have a clear definition of crime.

How Much Crime Is There?

Even with a clear definition of crime, we still have problems counting it up in the real world. In some cases, only the offender knows what he or she did. With other crimes, a victim knows but won't tell anybody. In still other cases, there seems to be no victim at all. Government officials and researchers have devised two main ways to measure crime. Thus, to be a student of crime, you must know the limitations in truly measuring crime.

The primary way crime is measured in the United States is through the Uniform Crime Reporting (UCR) Program, which began in 1930 and is conducted by the Federal Bureau of Investigation (FBI). The purpose of the UCR Program is to produce reliable, if not entirely accurate, information about crime reported to or discovered by law enforcement for the entire United States (FBI, 2008b). Because each state has slightly different criminal laws that change, the UCR Program provides national standards for the uniform classification of crimes and arrests (for further details, visit the FBI's Web site at http://www.fbi.gov/ucr). Notably, the UCR crime definitions are distinct and do not conform to federal or state laws, nor do they hold any legal standing. The program is voluntary, but more than 17,000 city, university and college, county, state, tribal, and federal law enforcement agencies provide information (FBI, 2008b).

Most crime statistics reported in the media are based on UCR data, which represent only a subset of crime that has actually occurred. The data consist of only those crimes reported to or found by the police for eight different types (Part I crimes)—murder, rape, robbery, aggravated assault, burglary, larceny-theft, motor vehicle theft, and arson—and arrests for all crimes. The FBI is even less strict in compiling information on Part II crimes, which comprise a vast variety of offenses, including simple assaults, drug crimes, alcohol offenses, miscellaneous sex crimes, running away from home, drunk driving, public drunkenness, disorderly conduct, and so on. Only the arrests that occur as a result of these crimes are counted.

The National Crime Victimization Survey (NCVS) measures crime in a different way, by asking individuals about victimization. It was developed in 1973 by researchers to measure victimization not captured by the UCR Program. Every 6 months, it samples about 100,000 U.S. residents who are more than 12 years old and asks them about their experiences with crime and victimization (BJS, 2009). The survey covers many of the same types of crimes as the UCR Program, but importantly, it asks about those that were never reported to police. From the NCVS, you will learn that violent crime is reported to police about 40% to 50% of the time and that property crime is reported about 30% to 40% of the time (BJS, 2009). However, the NCVS is not a perfect system either, as it asks only a small proportion of the population, does not track murder, and relies on citizens to recall their experiences and define the crimes themselves.

So, either method of crime measurement is problematic when trying to establish the true picture of crime. However, the UCR Program and the NCVS do attempt to consistently measure crime and victimization, which allows us to make comparisons across geographic areas and time, with some caution.

In recent years, other sources of crime data are giving us much more information. Businesses are getting better about reporting their losses to theft. Emergency room and other health data tell us more about violence. Interviews with youths tell us a good deal about their offending. In addition, data are improving vastly in other nations. The greatest crime improvements in North America are found in western Canada, where the Royal Canadian Mounted Police and Simon Fraser University have teamed up to organize police data on offenders and their crimes. Despite their good efforts, you should not take official reports as the "full truth" about the amount of crime in society.

But don't get too confused. We know a lot about measurement error, including systematic errors in measuring crime. There is simply no justification for continuing to treat the study of crime as a vague topic.

9. THE RANDOM CRIME FALLACY

Some people think that crime is likely to happen anywhere, anytime, and to anyone, and that it's a matter of time before you will be a victim, and a matter of luck when you are not. This *random crime fallacy* is pernicious because it eliminates personal responsibility and implies that crime cannot be prevented. Our purpose here is to show the opposite, that crime is both predictable and preventable. Crime is not just a random misfortune; it has patterns in time and place.

A wealth of research has found that crime indeed clusters in space and time. People and places that have been victimized in the past have a higher likelihood of being victimized again than do people and places that have never

been victimized (Farrell & Pease, 1993). In fact, the best predictor for victimization is if the person or place has been victimized in the past (Weisel, 2005).
Certain groups of people are disproportionately victims of crime—young
people, men of all ages, minorities, and people either divorced or separated
(BJS, 2009). Certain crimes cluster at particular times. For example, bar fights
tend to happen at the end of the night (Scott & Dedel, 2006), residential burglaries during the day (Weisel, 2002), and convenience store robberies at night
(Altizio & York, 2007). Finally, crime clusters geographically, with a large
amount of crime concentrated within a relatively few addresses (Sherman,
Gartin, & Buerger, 1989; Weisburd, 2005). Our lives are not random, so why
should crime be so? Indeed, our patterned behavior creates opportunities for
crime. Ridding yourself of the notion that crime is random is the first step in
understanding how crime clusters. As you proceed, you will understand more
and more about crime's clustering.

CONCLUSION

So many misconceptions have crept into your thinking about crime that you
must work to purge them. Statistics are thrown at you that don't paint the entire
crime picture. The media keep coming back at you with dramatic examples
that miss the point. The police and courts are important, but unrepresentative.
Defense mechanisms are strong for denying one's own crime potential. Young
faces continue to look kind and innocent. Victims remain in denial about how
easily they were outsmarted. Distorted images of crime organization and gangs
recur. Ignorant observers link crime to one pestilence after another, or fear the
most impractical occurrences. Those with axes to grind keep promising that
their agendas will stop crime. If you can push aside all of these distractions,
you are ready to break down crime into its most basic elements.

MAIN POINTS

1. The Dramatic Fallacy: The media distort crime for their own purposes, creating
   many of our erroneous conceptions about crime.

2. The Cops-and-Courts Fallacy: The importance and influence of police and
   courts as proactive controls over crime are overstated.

3. The Not-Me Fallacy: Crime is committed by everyone, and the "criminal" is not much different from us.

4. The Innocent-Youth Fallacy: Children are not the innocent bystanders to crime, but are overrepresented as offenders.

5. The Ingenuity Fallacy: Most crime is simple and most criminals are unskilled.

6. The Organized-Crime Fallacy: Criminal conspiracies and, specifically, juvenile gangs are attributed much greater organization and sophistication than they actually have.

7. The Agenda Fallacy: Crime is used haphazardly by a variety of people with moral, religious, social, and political agendas to support their causes.

8. The Vague-Boundary Fallacy: Crime can and should be defined so it can be studied across cultures and history while not becoming bogged down by opinion, different laws, and peculiarities.

9. The Random Crime Fallacy: Crime is not random but occurs in patterned ways that coincide with our routine behavior and everyday lives.

PROJECTS AND CHALLENGES

Interview project. Interview anyone who works in private security or retail trade. Find out what offenses are common and how they are carried out.

Media project. Take notes of three different nightly news programs. What crimes or crime statistics do they cover and how? What crimes do they fail to cover? How are the crime statistics represented?

Map project. Find an interactive crime mapping program of a city or police department and create a map of robberies or burglaries for 1 year. Note how they cluster. (For a list of interactive crime mapping programs, go to http://www.ojp .usdoj.gov/nij/maps/links.htm. For extra guidance in creating maps, see works by Boba, 2008; Chainey & Ratcliffe, 2005; Harries, 1999.)

Photo project. Take five plain photographs indicating that a crime might have been committed. Discuss.

Web project. Find some of the sources mentioned in this chapter, or their updates, via the Internet. Look at a table relevant to this chapter and describe it. You might find the Web page designed by the first author [MF] (crimeprevention.rutgers.edu) useful to you. Note in the appendix the list of important Web sites on crime and crime prevention.

NOTES

1. One author's [MF] brother, Ed, is a criminal lawyer who often handles juvenile cases. He said once, "I look at some of my young clients and tell myself, 'That's a kid.' Then I say to myself, 'That's also a criminal.'" Perhaps none of us can easily resolve this.

2. Drug gangs also exist, but we prefer to think of these as crime organizations in some sense. The word *gang* begins to lose its utility when stretched to cover too much.

TWO

CHEMISTRY FOR CRIME

The whole is more than the sum of its parts. In a school science class, you may have mixed baking soda, vinegar, and dish detergent, producing a small eruption. Mixes make surprises for crime, too, for people mix in different ways.

- Two "innocuous" individuals mix badly, getting into a fight.
- Two drug sellers compete, with one driving the other out of business, even away from crime.
- Two teenagers who behave well in the presence of parents get drunk and wild when together.
- A middle-aged man, conventional at work, gets caught somewhere else with his pants down.

Many people are volatile and reactive. Almost everyone has ups and downs, ins and outs, feelings of anger and calm, moods of conformity and defiance, and legal and illegal behavior. Emotional ups and downs are even more common for people who commit a lot of crime (Cogan, 1996). This is why you, as a student of crime, should pay attention to the specific circumstances affecting people at particular times, and how they act together.

THE SETTING

More than 40 years ago, Professor Roger Barker (1963) developed tools for studying these variations in how people behave together. He and his students studied thousands of details of daily life after dividing a small Kansas town into hundreds of "behavior settings," namely, slices of space and time. Barker noted exactly what people were doing, when and where, in what group sizes and what ages, in each of the settings studied.

These settings ranged from a school history class to guys hanging out on a street corner or at a gas station. Barker's students took special note of how each setting sets up individual choices. For example, the same individual would read in the library and play ball on the ball field—different activities in different settings. That town provided teenagers few settings for escaping adults for very long, but still helps us study youth processes in larger places and today's world (see Chapter 6). For each of us spends part of the day in some settings and part in other settings, affecting what we do ourselves and what happens to us.

Just as economists study markets, crime analysts can use settings. A setting is a location for *recurrent* behavior at known times (Felson, 2006, p. 102). A crime setting is where people converge or diverge—influencing their crime opportunities. Some settings are largely abandoned, except for the offender and target of the crime. On a street with few pedestrians, a robber can pick off a straggler. Other, busier streets make crime more difficult. Some settings invite people to get drunk or abuse drugs. Other settings are sobering. Clearly, settings help us understand a very physical process such as crime.

Moreover, one setting can set the stage for another nearby. This is well illustrated by the *ambush sequence*:

1. A public setting (e.g., a bar) is packed with strangers.

2. One of these strangers leaves, walking down a deserted street nearby.

3. A second stranger leaves the bar, following the first until there is nobody else around to interfere.

4. At the right moment, the second stranger attacks the first.

Dennis Roncek (1981) studied dangerous places, including bars that set the stage for nearby crime. He even found that living near high schools exposes people to risk of burglary and other crimes. P. L. Brantingham and P. J. Brantingham (1984, 1993, 1999) found that the tough bars in town were the likely origin of nearby crimes, including property crimes and violence alike (see also Scott & Dedel, 2006). Our point is that settings are the central organizing feature for everyday life. Most human activities occur in settings, and the features of some settings make them more risky for crime, whereas other settings are not very risky at all. To understand this, you should learn the stages of the crime sequence and the basic elements whose presence or absence make crime likely.

THE STAGES OF A CRIMINAL ACT

A crime incident can be meaningless without information about what led up to it. For example, one man punches another in the face and breaks his nose. But why did this happen? What led up to the event? What happened afterward? Your task as a student of crime is to think about the behavior sequences of everyday life. You should figure out when those sequences set the stage for crime. As you think about everyday sequences, you will better understand how the setting and the people in it get involved in crime.

You can divide the sequence for a criminal act into three stages (Felson, 2006, p. 42):

Stage 1. The *prelude*: The events that lead directly up to and into the criminal act, such as getting drunk, driving through a neighborhood, or waiting until no one is looking

Stage 2. The *incident*: The immediate criminal act, such as punching someone, breaking a window, or stealing a purse out of a car

Stage 3. The *aftermath*: Whatever happens after or as a result of the incident, such as the offender fleeing the scene, fencing stolen goods, or using a stolen credit card.

Notice that other criminal acts are often committed in the aftermath. Thus, a burglary sets the stage for a second crime, fencing stolen goods, and maybe a third crime and a fourth. Now that we've covered the sequence of a criminal act and its stages, let's dissect the criminal act into its elements.

THE ALMOST-ALWAYS
ELEMENTS OF A CRIMINAL ACT

Most criminal acts occur in a favorable setting. Criminal acts have three *almost-always elements*:

- A likely offender
- A suitable target
- The absence of a capable guardian against the offense

Starting with the likely offender, anybody might commit a crime. Yet the best candidate is a young male with a big mouth who does poorly in school, loses jobs, gets into traffic accidents, and ends up in the emergency room (Gottfredson & Hirschi, 1990). Daily life helps some people reach their full criminal potential, whereas others have a stunted criminal growth. Although most active offenders start young, some criminals may be late bloomers. The march of life provides new criminal opportunities, hence changing the pool of likely offenders as time goes on, while making some previous offenders more efficient or less so.

A suitable target is any person or thing that draws the offender toward a crime, whether a car that invites him to steal it, some money that he could easily take, somebody who provokes him into a fight, or somebody who looks like an easy purse-snatch.

"Guardians" should not be mistaken for police officers or security guards, who are very unlikely to be on the spot when a crime occurs. The most significant guardians in society are ordinary citizens going about their daily routines. Usually, you are the best guardian for your own property. Your friends and relatives also can serve as guardians for you and your property, as you can for theirs. Even strangers can serve as guardians if they are nearby, and the offender thinks they might turn him in or otherwise interfere with his plans. Some guardians are employees, such as a store clerk or other employee who helps protect against business crime.

With a guardian present, the offender avoids attempting to carry out an offense in the first place. *The guardian differs from the offender and target, because the* absence *of a guardian is what counts.* A guardian is not usually someone who brandishes a gun or threatens an offender with quick punishment, but rather someone whose mere presence serves as a gentle reminder that someone is looking.

THE CRIME TRIANGLE

You recall the not-me fallacy from the first chapter—the idea that "I'm too good for crime." Yet most people commit some crime sometimes. When will otherwise "good" people do bad things? The answer to that question has to do with temptations, opportunities, and controls. Professor John Eck of the University of Cincinnati devised the basic crime triangle (Exhibit 2.1), another tool for thinking about crime settings[1] (Center for Problem-Oriented Policing, 2009). You can always come back to that triangle if you get lost when studying crime. This triangle can be applied to anything from loud music to murder.

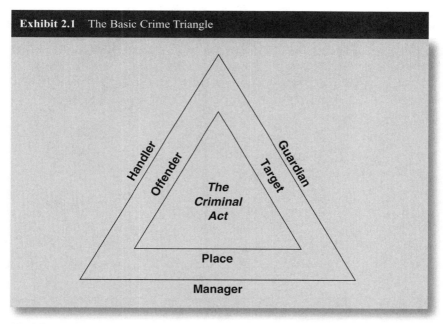

Exhibit 2.1 The Basic Crime Triangle

SOURCE: Designed by John Eck.

The small inner triangle in the center of the figure gives the three features of each crime problem: the offender, the crime target, and the place or setting where the crime occurs. These three elements need to converge in space and time.

But for that to happen, the offender must evade supervision of three types. The outer triangle tells us about this supervision. A handler supervises potential offenders; for example, mom, dad, the football coach, and the teachers are handlers whose presence or controls (e.g., curfew) discourages misbehavior. A place is supervised by a place manager, whose presence and alertness discourages crime from happening there. A target of crime within that place is any relevant person or piece of property. It can include money to steal, a person to attack, or somebody selling illegal drugs.

After evading the handler, the likely offender must find a place containing a crime target whose place manager is absent or indisposed. Then he must find a target with no guardian watching it. In sum, the offender moves away from handlers toward a place without a manager and a target without a guardian. This mix of divergences and convergences is summed up in the Dynamic Crime Triangle (Exhibit 2.2).

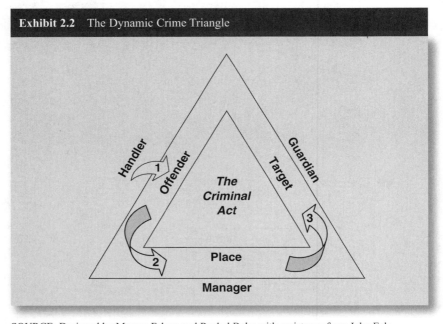

Exhibit 2.2   The Dynamic Crime Triangle

SOURCE: Designed by Marcus Felson and Rachel Boba with assistance from John Eck.

Pay close attention to place managers. Examples are hotel or store clerks and apartment building managers. They control the physical environment, set and enforce rules, oversee behavior, and thus influence crime. A school principal or mall manager might influence thousands of people and affect whether or not crime occurs. A bar owner with lax policies on underage drinking makes possible more drunk driving, fights, and sexual assaults in and around his bar. Managers are especially important for minimizing illegal drug sales and controlling public drinking.

Suppose two teenagers break into your car in the university lot, stealing your phone while you're in class. The teens escaped their parental supervision, found a car whose guardian (you) is absent. The parking manager is either absent or indisposed, and perhaps had not thought out well the design of lighting, video surveillance, or location of parking spaces.

THE OFTEN-IMPORTANT ELEMENTS OF A CRIMINAL ACT

These almost-always elements of a crime are supplemented by its *often-important elements*:

- Any props that help produce or prevent a crime, including weapons or tools
- Any camouflage that helps the offender avoid unwanted notice
- Any audience the offender wants either to impress or intimidate

Thus, an ideal crime provides the offender a target in the absence of a guardian against crime. A setting is even more ideal if it contains things that offenders like to use, audiences they wish to impress, and camouflage to hide their illegal action from anybody who might interfere. Additional features of daily life affect criminal acts, as you shall see in the progress of this book. Exhibit 2.3 depicts a setting containing almost-always and often-important elements of crime. The almost-always elements are shown in the upper half of the setting. The often-important elements are shown in the bottom half of the setting, where the camouflage hides the target, the props are accessible to the offender, and any audience can see the target and be seen by the offender. The general crime setting depicted in Exhibit 2.3 will, of course, vary among the great variety of criminal acts.

Notice in Exhibit 2.3 that the offender and target have entered the setting, and the guardian has exited. The offender may have props (such as tools to cut

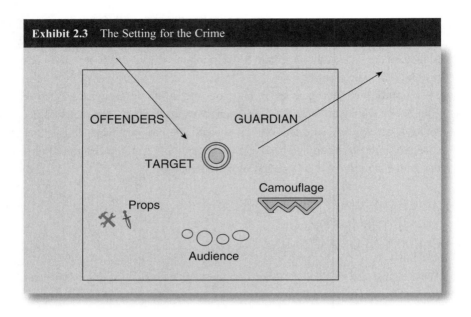

Exhibit 2.3   The Setting for the Crime

away what he's going to steal) and perhaps an audience (such as his other offenders, easily impressed by his mischievous display). And so it goes with everyday crime.

The direct-contact predatory crime predominates, even in a computerized age. True, the Internet and phone systems make many remote predatory crimes possible, but direct physical contact remains the most common criminal technique. That is why crimes are still so dependent on settings. This is true for each of three major categories of illegal action: predatory crimes, fights, and consensual criminal acts.

PREDATORY CRIMES

Although we often think of crime as predatory, with clear victims, in fact, ordinary crime also includes fights and consensual criminal acts.

Predatory crime occurs when the offender is clearly different from the victim, and the victim objects to the offender's actions. Notably, most Part I crimes are this type—rape, robbery, burglary, theft, auto theft. Other examples include child abuse, bullying, and identity theft. An ideal predatory offense requires at least one clearly guilty offender preying on at least one clearly innocent person or a property target the offender has no right to.

Predatory offenders are mostly impersonal. They do not care how the victim feels. Robbers just want the money, and if you give it to them politely, they usually just go away. There are exceptions. Some predatory attacks are personal, made by someone mad at someone else. For example, a local boy may get even with the neighbor who scolded him by vandalizing his car. But these noteworthy offenses do not negate the general rule that the offender is most oriented toward himself, not his victim. Indeed, we prefer the word "target" to "victim" because the offender is usually oriented toward either property or a person treated largely in physical terms.

The *predatory sequence* typically works like this:

1. A likely offender enters a setting.

2. A suitable target enters, too.

3. A guardian leaves.

4. An offender attacks the target.

Some settings are called *crime attractors* (Brantingham & Brantingham, 1998, 1999) because they are especially likely to draw offenders. In such cases, the first and second steps are reversed. In other settings, likely offenders and suitable targets are around for other purposes, but criminal acts emerge. These are called *crime generators.* Given enough time, settings may themselves transform, as shown in the following crime attraction sequence:

1. A setting has normal legal activities.

2. It assembles crime elements and becomes a crime generator.

3. Active offenders figure this out.

4. The setting becomes a *crime attractor*, with crime getting still worse.

You can readily see why the setting is central for crime analysis. That's where the action is. In most settings most of the time, crime does *not* occur. The question to ask is whether the almost-always and often-important elements are present. You can see that a setting is not fixed in time. It transforms itself, altering crime opportunities and outcomes. That process occurs not only for predatory crimes but also for other offenses.

FIGHTS

Fights are violent interactions involving two or more persons in the same conflict role. Most fights emerge from quarrels in which neither party is fully innocent. Often they are equally guilty, or almost so. They frequently have a pre-existing relationship, but sometimes strangers or near-strangers get into a fight.

True, one guy might have thrown the first punch, but if the other guy insulted or taunted him, it is hard to call that a predatory crime. Fighting words and a fighting response usually incriminate both sides. Typically, the police take the winner of the fight to jail and the loser to the hospital.

The *escalation sequence* typically works like this:

1. One party perceives an insult from the other.

2. He responds to the insult and escalates the confrontation.

3. That answer evokes a similar escalation.

4. Someone throws the first punch, and so it goes.

In a fight, the audience is especially important, for it enhances the embarrassment of being insulted and compels a response. If someone in the audience acts as peacemaker and face-saver, escalation can be averted. Comments like "Don't pay any attention to Joe" or "You guys knock it off" generally suffice to quiet the waters. On the other hand, some members of an audience may act as troublemakers or provokers. Statements such as "Are you going to let him say that to you?" tend to encourage escalation. These individuals are considered handlers because they are influencing the behavior of the offender(s).

Many social psychologists have investigated this escalation process in laboratories. The professor hires a "stooge," another student pretending to be a subject. An observer sits behind a one-way mirror watching. The undergraduate "subject" enters the room, thinking the experiment has not started. The stooge insults him, perhaps saying, "That's an ugly shirt you're wearing." The observer notes whether the subject responds to the insult and escalates the encounter, or shrugs it off. Some experiments introduce different audiences, or peacemakers, or provokers. The general picture of escalation given above is substantiated over many experiments (Kennedy & Forde, 1999; Tedeschi & Felson, 1994).

CONSENSUAL CRIME: ILLEGAL MARKETS

Consensual crimes involve at least two parties as illegal counterparts. For example, an illegal drug sale involves a buyer and a seller. A knowing buyer and seller of stolen goods also engage in consensual crimes.

Which markets are illegal may vary among nations and even American states, but they have a common pattern. The buyer and seller are cooperating. Neither wants to get caught. The offenders in an illegal sale, acting in different roles, are symbiotic in a biologist's sense: They depend on each other like flowers and bees.

Illegal markets for goods. These are for selling drugs or contraband, counterfeit money or products, as well as goods known to be stolen. As Chapter 5 will explain, many stolen goods are also sold without the knowledge that they are stolen. Illegal markets also include legal items sold illegally. Thus, the beer you might buy as an underage college freshman, even though from a legitimate store, fits the "illegal market" definition.

Illegal markets for services. Illegal markets may link buyers and sellers of prostitution, illegal sex shows, or even contract killings. A less sinister example is the ongoing chain of illegal services that makes New York's Chinatown work. The restaurants and their suppliers are mostly unlicensed and untaxed, and the delicious plate of noodles you can buy there at a cut rate reflects this.

Illegal markets for persons. These include slaves, women sold into prostitution, and persons smuggled into work arrangements across borders (see Chin, 1999, 2001; David, 2000; Zhang & Chin, 2002; Zhang, Chin, & Miller, 2007). Yet millions of people are smuggled into the United States with their own consent, and hired willingly by American employers.

Settings for illegal markets. These illegal transactions have physical requirements and depend on suitable settings. Exhibit 2.4 shows some of the settings for drug sales. Ideal for recruiting new customers are the outdoor, round-the-clock settings, such as the edges of parks or liquor stores. These are readily available to strangers who can spot what is happening quite easily. Next best are small apartment buildings or abandoned houses, but as these become well-known sources of drugs, offenders may move toward the previously described "ideal" places. Personal ties are less suitable for recruiting new users because the seller is (in the short run) limited to those he or she already knows. However, personal networks do fan out and can, in time, spread illegal sales to a broader audience.

Exhibit 2.4   Types of Drug Sale Settings and Recruiting New Customers

Ideal for recruitment	Might work well for recruitment	Inefficient for recruitment
Edge of parks	Small apartment buildings	Personal ties
Outside liquor stores	Abandoned houses	

Research on drug sales now pays very close attention to settings (Eck, 1995; Green, 1996; Jacobson, 1999; Natarajan, 2006; Natarajan & Hough, 2000; Rengert, 1996; Rengert, Chakravorty, Bole, & Henderson, 2000; Reuter, 1998). A remarkable private entrepreneur, Severin Sorensen, has achieved major reductions in outdoor drug sales and related crime by redesigning settings where drugs are sold. Police crackdowns on the ideal settings for selling drugs serve to reduce the problem greatly (Knutsson, 2000; see also Rengert, 1996; Weisburd & Green, 1995). After such settings are denied, the population of drug abusers tends to get older and older. It is short-sighted for police or other officials to say, "Those drug sellers just went indoors; I'm giving up." The outdoor settings provide the best *recruitment of new drug abusers,* hence replenishing a sick and aging clientele. Denying the best settings, as Rengert et al. (2000) explain, serves a serious purpose.

CALMING THE WATERS AND LOOKING AFTER PLACES

Several roles have already been described for people who calm the crime problem. Guardians can prevent property crime and illegal sales. Handlers or peacemakers in particular can discourage fights. Managers also serve to prevent crime settings from deteriorating because they oversee and monitor the settings:

- Homeowners and long-time renters
- Building superintendents, doormen, and receptionists
- Bartenders, managers, and owners
- Small-business persons and store managers
- Street vendors
- Security people with focused responsibilities
- Park and playground supervisors
- Train station managers
- Bus drivers

Each place manager has an incentive to prevent crime. The owners and long-time residents want the place to be safe. The businesspeople want to make money. The others want to keep their jobs and be safe, too. Society has other guardians, but place managers are probably the most important of all. On the other hand, they cannot watch everything everywhere.

Crime analysts have also paid a good deal of attention to natural surveillance. That includes ordinary citizens going about their daily lives, but providing by their presence some degree of security. We can learn a great deal from Oscar Newman's (1972) classic distinction:

1. Private space,

2. Semiprivate space,

3. Semipublic space, and

4. Public space.

Private space might include the inside of your home. The area just outside your home—such as an apartment lobby—is probably semiprivate. That means you have a good deal of control over who goes there. Semipublic space is farther out, such as a yard in front of an apartment building, where the public might not always go but can get to rather easily. Public space, such as a street or sidewalk, is especially difficult for you to control, and hence provides the least security. But you can never be entirely sure of keeping any space secure from crime. That's why you must pay close attention to crime targets themselves.

RISKY SETTINGS

Daily life is divided into different types of settings, some of which generate a lot of crime (see Clarke & Eck, 2005, Step 15; Clarke & Eck, 2007; Felson, 1987). The riskiest types of settings are

a. *Public routes*, especially foot paths, parking facilities, and unsupervised transit area

b. *Recreation settings*, especially bars and some parks

c. *Public transport*, especially stations and their vicinities

d. *Retail stores*, especially for shoplifting

e. *Residential settings*, especially for burglary and theft

f. *Educational settings*, especially on their edges

g. *Offices*, especially when easily entered for theft

h. *Human services*, especially hospitals with 24-hour activities

i. *Industrial locations*, especially warehouses with attractive goods.

HOT PRODUCTS

Some products are stolen much more than others. Start with cars. The Highway Loss Data Institute (http://www.carsafety.org) shows that specific makes and models differ greatly in their probability of being stolen. Exhibit 2.5

Exhibit 2.5 Theft Losses for Two-Door Cars, 2003–2005 Models, United States

Model	Theft Index*
Honda Civic Hatchback	236
Chevrolet Monte Carlo	156
Acura RSX	146
Mitsubishi Eclipse	141
Honda Accord	112
Pontiac Grand Am	111
Dodge Stratus	111
Hyundai Tiburon	102
Toyota Celica	91
Volkswagen Golf	90
Honda Civic Coupe	86
Chrysler Sebring Convertible	66
Chevrolet Cavalier	66
Scion tC	63
Toyota Camry Solara	56
Ford Focus	49
Pontiac Sunfire	49
Audi A4 Cabriolet Convertible	48
Volkswagen New Beetle Convertible	42
Saturn ION Quad Coupe	32
Volkswagen New Beetle	28
Hyundai Accent	27
Mini Cooper	24

SOURCE: Highway Loss Data Institute, www.carsafety.org, September 2006. Electronic copy available at http://www.iihs.org/research/hldi/ictl_pdf/ictl_0906.pdf.

*Calculated from average loss payments per insured vehicle-year, then indexed to the average for all two-door cars. Thus, 104 is 4% worse than average.

sums up the theft loss data for 2003 to 2005 two-door car models, as reported in September 2006. It shows that the Honda Civic hatchback generates *9.8 times as much theft risk* as the Mini Cooper. The Honda Accord creates about twice as much insurance theft problem as the Chevrolet Cavalier (see Brown & Billing, 1996; Clarke, 1999). Even trucks vary greatly in their ability to attract thieves (Brown, 1995). Clearly, some products attract the attention of thieves much more than others. Jewelry, cash, and small electronic goods are more likely to be stolen in burglaries than most items in the house.

According to the National Retail Security Survey, in recent years, those products most stolen from stores have been magazines, shirts, jeans, items with the Hilfiger or Polo label, CDs, videocassettes, beauty aids, cigarettes, batteries, condoms, sneakers, Nike shoes, hand tools, jewelry, and action toys (Hayes, 1997b, 1999, 2000, 2003). As times change, types of products stolen change as well; we are seeing iPods, DVDs, and video games being stolen often today.

Past studies have compared retail value per pound of most stolen goods (see Cohen & Felson, 1979; Felson & Cohen, 1980). Washing machines are worth several hundred dollars, but only about $5 per pound. They are not usually stolen. In contrast, electronic consumer goods are high in value per pound and attractive targets of crime. Clarke's (1999) pamphlet *Hot Products* tells us a good deal about which items are most likely to be stolen. That and other sources help us understand an important issue in the chemistry of crime. What features of crime targets make them so inviting?

Items That Invite Theft

The property offender wants to steal something valuable, enjoyable, and available. He wants to be sure he can remove it from the property. As he removes it, he wants to be able to conceal it. After he removes it, he wants to have a way to dispose of it, perhaps selling it to somebody else.

Clarke (1999) rearranged those words to form this acronym for goods "craved" by thieves. Hot products tend to be

Concealable

Removable

Available

Valuable

Enjoyable

Disposable

Cash solves all six problems, so it will be a hot product as long as we are still printing it. Jewelry is often desirable, too, for it is easily concealed on one's person. A very common car is easier to conceal than an unusual model (such as a Rolls Royce) that is easy to spot. Small items and those on wheels are the easiest to remove. Light televisions are more vulnerable than the heaviest models. If thieves had to carry the car on their backs, there would be no car theft.

Items produced in great numbers and placed near a door are most likely to be stolen because they are available (Clarke, 2002). Valued items to young thieves might be popular CDs, but not Beethoven. Old-style maternal minivans are almost never stolen because young car thieves do not want them. Offenders are more likely to steal products that are fun to own or consume. They tend to enjoy alcohol, tobacco, expensive clothes, fancy shoes, DVDs, or condoms. Thieves also prefer items easy to dispose of (see Chapter 5).

Hot Products Are Affected by Their Settings

Hot products are affected by the setting that contains them (see Chapters 4 and 9). A household burglar targets both the home and the goods inside it. Many burglars assume that something good is inside, paying attention to the building in making their decisions. The offender considers tall bushes that could conceal entry and exit and may take into account the proximity to the street and driveway, enabling removal of what is taken. The offender also considers the environment and whether people are in the building, which is why residential burglaries happen predominantly during the day when the residents aren't home. Subsequent chapters of this book will go into more detail on how settings affect offender decisions, which also then provide tools for preventing crime.

Stores that hide their cash-counting operations in an upstairs office gain some protection by not tempting people needlessly. Being out of sight also helps protect some illegal transactions from prying eyes. Access is the ability of an offender to get to the target and then away from the scene of a crime.

With enough effort and sufficient tools, every offender can get to every target, but the whole idea of crime is to get things the easy way, acquiring few skills and applying little effort over a short time. Easy access is essential for ordinary crime. Many offenders are probably even more concerned with their exits than their entrance, wanting a safe getaway. Flashing money, leaving valuables out, making obvious the lack of locks, putting valuables by the front window where people pass by, living on a busy street, inviting too many people to see your new stereo, putting a shiny new car on the street, letting the hedges overrun the back windows—all of these can generate extra crime risk.

Targets Vary by Offender Motive

Legal definitions of criminal acts often hide human differences. For example, auto thieves differ in their motives and hence in the targets they pick (Clarke & Harris, 1992):

- A joyrider takes a trip for fun, picking a flashy and fun car.
- A "traveler" chooses almost any car that's convenient and drives home.
- A felon steals a car to perform another crime, picking a fast model.
- A parts chopper selects a common model 3 to 5 years old for which parts are in demand.
- A "shipper" takes a luxury car to sell abroad.

As motives shift, so do targets.

When Heavy Items Are Stolen

We noted earlier that washing machines and other heavy appliances are worth a lot of money but too heavy for the average thief to remove quickly. But the exceptions illustrate how relative these concepts really are. Washers, dryers, dishwashers, and similar appliances are sometimes burgled from rural homes, vacation homes in the off-season, or construction sites. In these three cases, pickup trucks are probably common, making the burglary easier to conduct and less likely to stand out.

As a general rule, the weight of items burgled increases as one goes farther from the city center. Inside the city, burglars are more likely to work on foot, doing their best to carry off money or jewels. In suburbs, more burglars use cars and can remove televisions or other electronic goods. In rural areas, with pickup trucks at hand and guardians away, offenders may find it easier to load up with whatever they want. At construction sites, burglars may look and act like construction workers and have the equipment and trucks to uninstall and drive away with the appliances (Boba & Santos, 2008; Clarke & Goldstein, 2002).

Heavier items are also stolen when wheels provide a getaway. Cars, motorcycles, mopeds, and bicycles face high risk. A family returned from vacation to find that someone had broken into their garage, taking the two cheaper bikes but leaving the expensive one that had no air in the tires. On some college campuses, bicycles are often stolen. In nations such as the Netherlands and China, where bicycles are a primary form of transportation, their theft is endemic.

Theft Trends

To learn which targets offenders crave, find out what is popular among youths. What brands of shoes are in style? What video games are in demand? What clothing labels are in current fashion? What items are getting lighter? Has the packaging changed? Are items being stored where they are more subject to theft?

Changes in goods and money are very important for crime rate trends. Cohen and Felson (1979) have shown that one of the major causes of the mushrooming crime rates in the United States after 1963 was the proliferation of lightweight durable goods that were easy to steal. Knutsson and Kuhlhorn (1997) in Sweden and Tremblay (1986) in Quebec, Canada, demonstrated that easier use of checks and credit cards without careful identification produced a proliferation of fraud. Americans using far less cash helped to produce the declining U.S. crime rates since 1990 (Felson, 1998). You can see that the CRAVED model tells us a great deal about theft variations and trends. These practical features also apply to targets of violent crime and how offenders select these targets.

CRAVING VIOLENT TARGETS

Like property offenders, those who commit violence also have problems to solve. The CRAVED model applies well, especially if we shift to verbs.

A violent offender generally needs to *conceal* the violent act, as well as steps before and after it. He must *remove* himself safely from the scene; *avail* himself of a convenient human target for violent attack; find a target of *value* in his own mind; *enjoy* the criminal act, or at least avoid pain to himself; and *dispose of* incriminating evidence, even the victim.

To solve these problems, offenders pick dark and isolated alleys or parks, spots with easy entry and exit for themselves and a good chance to ambush someone, victims who are weak and alone, and victims who give pleasure and cannot easily inflict pain; they even consider simple ways to dispose of weapons. Kidnappers, for instance, probably wait until the victim is most available. They may pick a victim from a wealthy family. They may enjoy getting even with that family or someone in it. And they may find a remote location that permits disposal of their victim without being seen. A blitz rape (LeBeau, 1987) is also a highly physical crime. An example is when an offender grabs a victim from a dark street and then rapes her in nearby bushes. The target and setting are concealed. The target is usually small and weak enough to be removed from one spot to another. The offender may live, work, shop, or prefer recreation in certain areas where he finds a victim available. Rapists normally are attracted to victims of younger ages, reflecting their pursuit of pleasure (R. Felson, 2002; Tedeschi & Felson, 1994; for some reason, this point is controversial). Some rapists even kill victims and dispose of the bodies, or at least try to dispose of any evidence of their own criminal behavior. Store robbers attack cash registers that are concealed by advertisements plastering the windows. They seek removable cash near the door and available stores near the freeway or large boulevards. These robbers would generally prefer stores that have no drop safe and whose employees are, in some sense, compliant. Stores with lots of small bills assist the offender's disposal of the loot. Moreover, those offenders who use a mask or disguise need a place to dump it unseen.

The literature increasingly recognizes the physical side of assaults as more than incidental. Richard Felson (1996) explained in *Big People Hit Little People* how physical size and strength influence assaults. The classic book by Olweus (1978), *Aggression in the Schools: Bullies and Whipping Boys,* explained how larger and stronger children abused those smaller and weaker. Studying Norway and learning from other nations not known for their violence, Olweus showed that violence between children is probably universal among nations and very common in schools (see also Baldry & Farrington, 2007; Espelage & Swearer, 2004; Greif & Furlong, 2006; Smith et al., 1999).

Bullying also applies to other ages and settings, as explained in Randall's (1997) book *Adult Bullies* and Barron's (2000) work on "workplace bullies."

Those who crave sexual release with children or young teenagers can also be pragmatic. James M. Mannon (1997) considers how easy it is for stepparents to gain access to stepchildren, priests to choirboys, and child care workers to young children. Offenses usually occur in settings with guardians absent and unable to intervene. Family members are also more vulnerable to intimate victimization when isolated. Baumgartner (1993) found that wives with the support of family members have less risk of being beaten by husbands (see also Davis, Maxwell, & Taylor, 2006; Felson, Ackerman, & Yeon, 2003; Felson & Outlaw, 2007; Zolotor & Runyan, 2006). Opportunities for sexual coercion also seem to be linked to physical access (Tedeschi & Felson, 1994; Wortley & Smallbone, 2006).

The essential point is that both property and violent offenses are physical acts involving tangible targets whose specific features make them more suitable for offenders to attack.

THE GENERAL CHEMISTRY OF CRIME

Each crime type has its particular chemistry. Crimes also have a common chemistry. For each setting, consider its presences and absences, its entries and exits, and how these make a particular crime likely to occur. Even solo drug abuse depends on the absence of anyone to interfere. Some settings favor one offense but not another; for example, a crowd drives away most offenders but invites pickpocketing. Some settings enhance many types of crime. Scholars have shown that blocks containing bars or high schools have more crime (Roncek & Lobosco, 1983; Roncek & Maier, 1991). Other work uses the medical term "hot spots" to show that a small number of addresses (such as a large retail store or a tough bar) generate many more than their share of calls for police service (Sherman et al., 1989). Even within a high-crime area, most specific addresses appear to be quite safe, whereas a few addresses generate most of the problem.

To understand where many types of crime risk occur, Brantingham and Brantingham (1998, 1999) offer three terms:

1. *Nodes:* settings such as homes, schools, workplaces, shopping or strip malls, and entertainment areas. They provide particular crime

opportunities and risks. A node that favors one type of crime might not favor another, but specific crime risks differ greatly among nodes.

2. *Paths:* leading from one node to another, also offering crime opportunities and risks. Not only do paths conduct more people per square foot—hence providing offenders, targets, and guardians—but also lead people to nodes that might involve them, one way or another, in crime.

3. *Edges:* places where two local areas touch. Crime is often most risky here. At the edges of an area, outsiders can intrude quickly and then leave without being stopped or even noticed. For example, college students might find their cars broken into when they park at the edge of campus. Those who can only find parking at the edge of a high-crime area suffer greater property or even personal risk.

Finally, some human categories face the greatest general risk of crime victimization, including young males, especially if they are single, living alone, drinking a lot, and staying out late. Those who are past victims or are themselves frequent offenders also face extra victimization hazards (Hindelang, Gottfredson, & Garafolo, 1978; see also Fattah, 1991; Kennedy & Forde, 1990; Lasley & Rosenbaum, 1988). If you put together hot products, young males, people living alone, heavy drinking, late hours—at risky nodes or along edges of areas—you have created the ideal chemistry for crime. But this is just a generalization; in this book, you will learn details that vary from that theme.

CONCLUSION

As you read this book, you should look at life from the offender's point of view. The offender sees targets and considers whether guardians are absent. The offender picks suitable settings from his own point of view. The offender is the one who takes offense, gets in a quarrel, and then gets into a fight, so his viewpoint is the one you need to consider to understand how that happens.

Yet offenders are but one element in a crime, and probably not even the most important. Predatory crimes need targets with guardians absent. Fights thrive on audiences and troublemakers without peacemakers. Illegal sales depend on settings where buyers and sellers can converge with camouflage present and place managers absent. Offenses depend on quick entry and exit.

Targets most craved by offenders are concealable, removable, available, valuable, enjoyable, and disposable. We shall argue in the course of this book that *opportunity is a root cause of crime* (Felson & Clarke, 1998). Everyday life tempts and impairs potential offenders, influencing their decisions about crime, as the next chapter considers in detail.

MAIN POINTS

1. A setting is a location for recurrent use, for a particular area, at known times; a crime setting is where people converge or diverge to influence crime opportunities.

2. The sequence of a criminal act can be broken down into three stages: the "prelude," the "incident," and the "aftermath."

3. A criminal act has three elements almost always present: a likely offender, a suitable personal or property target, and the absence of a capable guardian against a crime.

4. The crime triangle illustrates the relationships among offender, target, place, and time and mechanisms that can influence crime opportunities, handlers, guardians, and managers.

5. Predatory crime occurs when the offender is clearly different from the victim, and the victim objects to the offender's actions.

6. A fight begins with an insult and then escalates, often affected by an audience and alcohol. Peacemakers can quiet things down and prevent that escalation.

7. Illegal markets are those in which participation is voluntary and consensual. Types include illegal markets for goods, services, and persons.

8. Place managers have a large span of control, making their role in crime opportunities important.

9. Settings, some of which are at higher risk for crime, include public routes, recreation settings, public transport, retail stores, residential settings, educational settings, offices, human services, and industrial locations.

10. Offenders "crave" hot products and targets of violence for specific reasons.

11. Crime risks occur at nodes, paths, and edges.

12. Opportunity is the root cause of crime.

PROJECTS AND CHALLENGES

Interview projects. (a) Interview any retail store owner, manager, clerk, or someone who worked in a retail store about what gets shoplifted. If you yourself worked in a store, supplement the interview with your own experience. (b) Interview a bartender, bar owner, or server about conflicts—how they develop and are prevented in a bar setting, and how to keep them from escalating.

Media project. Find a rich newspaper account of any plain crime and describe the prelude, the crime incident, and the aftermath. Draw from this chapter to make sense of the account.

Map project. Draw (a) the floor plan of a retail store and consider the vulnerable places for shoplifting based on layout and types of goods, or (b) the inside of two bars and consider which one is more likely to have people bumping into each other and other conflicts.

Photo project. Find four different places on campus most suitable for a crime to occur. Photograph each one and then describe it.

Web project. Use the Internet to gather information that helps you elaborate any point in this chapter. You might find the Web page one author [MF] has designed (crimeprevention.rutgers.edu) useful to you.

NOTE

1. Also called the "problem analysis triangle."

THREE

CRIME DECISIONS

---•◦•---

C rime is crass, and crime analysis must be, too. You must figure out how the offender thinks and decides, what he considers or ignores.

THE DECISION TO COMMIT A CRIME

Jeremy Bentham, born in 1748, figured out a good deal about how people make decisions. His *utilitarian model* gets us started in understanding why people choose to commit illegal acts today (see Bentham, 1789/1907). By this model, every individual seeks to gain pleasure and avoid pain.

To understand criminals, modify Bentham's ideas a bit. Nobel Laureate Herbert Simon (1957) taught us the principle of *limited rationality*. The average citizen can keep in mind only about three to five specific goals when making a decision, usually fewer. Thus, people make decisions, taking into account some costs and benefits, but they usually don't think of everything.

You can learn something about this process from musicians and athletes. Jazz musician Duke Ellington (1976) explained that "improvisation" is not spontaneous. Rather, he planned the music a moment before playing it. Athletes also make quick decisions, such as whether to swing at a baseball, or whether to kick the football this way or that. The point is that people can think quickly before they act.

Offenders make quick choices. Like athletes at a game, or drivers on the freeway, these choices are not fully spontaneous. Based on limited rationality,

Cornish and Clarke (1986) have modernized our notion of how offenders think and decide. Offenders are not much different from everybody else. Those who commit a crime follow roughly the same decision-making processes as those who make economic or musical decisions. Offenders are neither too careful nor totally spontaneous. They think a little, but not for long. You can use this in trying to understand crime.

The offender seeks to gain quick pleasure and avoid imminent pain. Offenders make decisions that are concrete—depending on specific setting, offense, and motive—while staying out of immediate trouble. A decision made a split second before acting is still a decision. Imagine a countdown: 50 seconds before, head toward the store; 20 seconds before, look around the parking lot; 10 seconds before, enter and look around the store; 2 seconds before, pull out the gun; 2 seconds after, leave with the money.

How Cautious? How Casual?

Most criminals take a rather casual approach to crime, still making decisions. The point of crime is to get things without having to work hard and without much dedication; thus, most crime is quick and easy, and most offenders are unskilled. That does not mean they are stupid, merely that they do not usually put forth a lot of effort.

Nor are they extremely daring. Of course, "daring" is a relative term. Most active offenders are daring enough to break the law and to risk serious personal consequences. But they usually go for the easy pickings. They are daring in comparison to nonoffenders but usually avoid the worst risks. Yet when an offender thinks an offense will be extremely rewarding, he might take a strong drink and go for it. In other words, the most active offenders are daring compared to nonoffenders, but they still avoid the worst risks.

How Much Do Offenders Consider?

The ordinary robber is an important example. Martin Gill (2000a) gathered some of the best information we have from British commercial robbers. Gill asked 15 basic questions to assess how careful they had been in robbing particular banks, stores, post offices, or other establishments. He asked offenders—among other things—whether they had visited the target earlier, kept the target under surveillance, worn a disguise, chosen the specific day and time, considered its location, and paid attention to its security measures. Most commercial robbers had taken a few of these steps, at most.

Consistent with Gill's British robbers, Feeney's (1986) interviews with California bank robbers found that most did not make intricate plans and had never been in the bank prior to robbing it. One in five got into "accidental robberies" resulting from burglaries, fights, or something else (such as a sudden impulse). Exhibit 3.1 illustrates how little planning these robbers did. Only 3 of 112 robbers reported doing a lot of planning. Feeney gives examples of offenders driving along with passengers who were totally unaware of what was about to happen. Many a robber even surprises himself.

This does not mean, however, that robbers do not make any decisions. It merely shows that they do not act with great care, or need to. This general view of robbery is confirmed by other studies (Altizio & York, 2007; Bellamy, 1996; Jacobs & Wright, 1999; Morrison & O'Donnell, 1996; Petrosino & Brensilber, 2003; Scott, 2001; Wright & Decker, 1997). So we are left with offenders who think a bit about a few things, avoiding the worst risks.

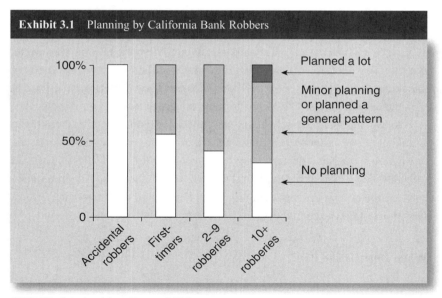

Exhibit 3.1 Planning by California Bank Robbers

SOURCE: Calculated from Feeney (1986).

Offenders Are Attuned to Certain Details

If you interview an offender, he may show bravado. If you ask him about a particular crime, he won't want to self-incriminate. He may feed you a sob story in hopes of getting out of jail. He may deny guilt or claim a bad childhood. And it may be true, but it won't get you any closer to knowing about crime. The best way to get a straight answer is to ask the offender specifically how a crime is done—exactly how to break into a car, shoplift a store, or rob a bank.

Paul Cromwell and his colleagues (Cromwell, 2003; Cromwell, Olson, & Avary, 1991, 1993) learned a great deal about the modus operandi for burglary by simply driving around with local burglars, peppering them with questions on the spot, giving them no time to make things up (see also Hearnden & Magill, 2004; Schneider, 2003). The researchers asked questions that could be answered without fluff:

- Why do you pick this street?
- Would you break into this house? Why not that one?
- Would you go back later?

Most of these burglars said they wanted to break into low-risk homes that were easy to enter. They start by looking to see if anybody is in the house and the ones next to it. Three clearly empty houses in a row put the middle house at greater risk because the neighbors are not there to prevent the crime. Burglars first look for activity, then they might probe by ringing doorbells or knocking on the door to see if anybody is there. They drive around a little to find the easy pickings. The idea is to look and act quickly.

As we noted earlier, an offense can be *almost* spontaneous without being irrational. Even without real planning, an offender responds to *cues in the immediate setting* and decides what to do (see Wortley, 1997, 1998). Even an offender with a plan may discard it if the cues tell him to do so. Environmental cues are the crucial link between individual choice and the immediate settings that impel or constrain choices.

When Does Crime Pay?

It's easy to declare, "Crime doesn't pay." More broadly, we might ask, "Does crime reward the offender?" The answer to that question depends on the time frame. Illicit sex usually gives a quick orgasm, but it might set in motion problems for the person doing it. And so it goes with other transgressions.

Most offenses pay in the very short term, namely, on the day they are committed. In the long run, most offenders suffer enough bad experiences to justify our conclusion that crime eventually fails them. Bodies deteriorate, other people counterattack, and one is left with a bad resume for later life. That ought to provide sufficient motivation to go clean. *Offenders almost certainly suffer more from the consequences of their own lifestyles than from the actions of public officials.*

On the one hand, Wilson and Abrahamse (1992) calculate that crime does not pay over time. On the other hand, recent work by Tremblay and Morselli (2000) disputes their arithmetic, concluding that crime often pays satisfactory dividends for a couple of years (see also Morselli & Tremblay, 2004; Morselli, Tremblay, & McCarthy, 2006; Tremblay & Pare, 2003). You can see why many people stick with crime for some time before they pay the price or have to quit.

THE DECISION-ENVIRONMENT PARADOX

An important paradox is that

- The everyday environment *external* to offenders guides their illegal acts; yet
- Offenders make a calculated choice.

If the first statement is true, how could we blame, convict, and punish persons not acting entirely by free choice? Given the second statement, how can we justify our claims in the first statement?

We believe that both (a) and (b) are true. To resolve the paradox, understand that the everyday environment provides cues *external* to the offender, guiding her illegal acts. But she still makes decisions in response to those cues. Not everybody responds exactly the same to any given environmental cue.

Indeed, everyday stimuli can entice offenders to commit a criminal act at a particular time or place, or might push them away from such offenses. But these stimuli are not perfect, nor do people respond like machines or like very primitive life forms. Rather, people vary in what they do. Some people take advantage of a very easy crime target, but shy away from any other target. Others take more of a chance.

Moreover, everyday settings vary moment to moment in their degree of temptation or control—the cues they emit—hence the degree of choice they provide. Constraints on individuals shift quickly as events unfold. The next sections explore these variations and shifts, even within a very short time span.

Settings Offer Different Choices at Different Times

To understand how temptation and controls shift so quickly, consider the *cue-decision sequence*:

1. Someone enters a setting

2. **containing some cues that transmit temptations and controls;**

3. after quickly noting and interpreting these cues,

4. he or she decides whether to commit a criminal act.

We have included the second step in the sequence in boldface type to emphasize that settings vary in the cues they emit. These cues often come from or are created by guardians, managers, and handlers discussed in Chapter 2. Some settings are overbearing in their controls, whereas others have hardly any at all. For example, in some settings, several people are watching their property, making it difficult for the offender to steal it, whereas in other settings, nobody is around, presenting little control to the would-be offender.

Yet temptations are themselves very uneven. Some settings have strong temptations, such as money lying around; others have nothing tempting whatever. In other words, some times and places offer an individual both temptation and room to resist it. Other times and places offer either little temptation or plenty of temptation but not much chance to succumb to it with impunity.

We can only evaluate personal responsibility with respect to a particular setting in which an individual is contained, taking into account the temptations and controls it presents to that individual. At 10:00 a.m. on Thursday, you might be walking into a setting that tests you with tremendous temptation and offers few controls, whereas your friend might be sitting with no temptation, and a third person faces temptations but strong controls. If you resist temptation, you deserve greater credit than your friend or the other person does because neither has to deal with it. However, from time to time, every human being is tempted and everybody is controlled, making decisions accordingly.

Options Shift

Richard Felson[1] describes the *potato chip principle*: No one can eat only one potato chip. The key decision, therefore, is to avoid eating the first one (see Baumeister, Heatherton, & Tice, 1994; Henry, Caspi, & Moffitt, 1999; Webster, 2005). The principle tells us that blame and control may vary as a person goes through even a few minutes of life. This applies to crime in very serious ways. Consider the *disinhibition sequence*:

1. A young man drinks some beer with friends.

2. He gets "high" and drinks still more.

3. Then he smokes some marijuana, getting "higher."

4. Some of the boys commit a burglary.

An early decision sets the stage for what happens later. One drink leads to another. A misdemeanor can lead to a felony (see Chapter 8). It may get harder and harder to pull away from and resist trouble. But if you ever get arrested for burglary, do not tell the judge to let you go because you were intoxicated. The legal system blames you anyway. The judge probably will get mad and tell you, "If you can't handle alcohol, don't start drinking it." To quote the stern but astute father, "You shouldn't have been there in the first place. If you had listened to what I told you, you wouldn't be in trouble now." Just as that father focuses on the points of maximum self-control to assign blame, the son emphasizes the moments of minimum self-control to escape blame. You need to look at both points in order to understand crime fully.

We all could make a decision that nearly enslaves us for some time afterward. Then a flash of freedom arises, a crucial juncture for the next decision. Moving through life, a person never has complete freedom or complete constraint, but the degree of constraint shifts by time, place, and setting. Even in the course of a day, one has more pressures and constraints here and fewer there. For example, when home with parents, one is pressured against getting drunk; while out with peers, one's pressures may go in the opposite direction. Yet a person still deserves blame for picking bad friends, placing oneself in dangerous settings, or abusing substances. That's why "pre-criminal situations" (Cusson, 1993) are so important for crime, as well as for drawing blame for bad outcomes. Like photons shifting into matter, blame takes quantum leaps in the course of everyday life.

Cues Are Needed to Assist Self-Control

We are all born weak, but parents and teachers try to teach us self-control to help us resist various temptations; to keep us studying or working, out of the kitchen, or away from the bottle; to help keep our mouths shut when the boss, customer, spouse, or teacher says something that tempts a nasty reply; to avoid fights, drugs, and thefts; and to keep doing our homework.

Overwhelming evidence shows that those low in self-control are likely not only to commit crimes but also to have many problems, including traffic accidents; time in the emergency room; crime victimization; excess smoking and drinking; and making a mess of school, work, and family relationships (Britt & Gottfredson, 2003; Gottfredson & Hirschi, 1990; Hirschi & Gottfredson, 1993a, 1993b, 1994, 2000; Junger & Marshall, 1997; Pratt & Cullen, 2000). Self-control also applies to paying attention; setting goals; managing money; avoiding procrastination; and regulating one's own impulses, appetites, moods, and thought processes (Baumeister et al., 1994). Those low in control have trouble regulating their minds, mouths, or actions. They might blurt out improper ideas to an interviewer, even without belonging to a special subculture.

If many people are less controlled, that does not mean they have zero control. They might still respond to immediate threats, to cues that say "not now." Their short attention spans do not bar a quick glance to see who is looking.

Those With Greater Self-Control Still Need Extra Reminders

It is a mistake to think of self-control as a pure individual trait that carries you through the day no matter what. It makes more sense to talk about *assisted self-control,* with everyday life helping people along. Most of us need *reminders* to follow the rules. A reminder is nothing more than a cue designed for people who have a good deal of self-control and are basically inclined to follow the law. But they need a bit of help.

Society delivers these reminders in various settings. That keeps the offending population smaller and the others at bay. Reminders come in verbal and physical forms. Verbal reminders have a half-life, that is, a decay period. A parent's reminder to a teenage son to take out the garbage cans may have a short half-life, requiring repetition a minute later. A police officer's reminder to turn the music down could have a half-life of only a couple of minutes with some people, although others need to be told only once and then will remember for years to come. Environmental reminders are often far better, for they keep "telling" people not to try something, even when nobody is on duty (see Chapter 9).

Stigmas Are a Poor Substitute for Environmental Cues

In a village, people tend to know who steals things. They can keep an eye on these individuals and make their stigma work against crime. In metropolitan America, we usually know little about the criminal records of neighbors; juvenile records are sealed, and unless the media report it, we seldom know about most other offenders. So people use *careless stigmas* based on looks, guesses, skin color, or sloppy rumors. It's inefficient to stigmatize a whole race or other grouping: The correlations are too small.

Stigmas are a poor substitute for environmental cues, which remind *everyone* not to try an illegal act. Stigmas encourage the community to build a wall against those they think are the criminal race or group, while letting the others do whatever they want. Yet the *local* boys are the ones who do most of the local crime. Insiders notice targets while avoiding scrutiny themselves (Brantingham & Brantingham, 1998, 1999). Wrongly assuming that outsiders are their problem, residents of middle-class areas may leave insiders without scrutiny.

Moreover, by refusing to enter lower-class neighborhoods or to insist on enforcement there, middle-class people are allowing crime to happen. They often continue in this mind-set until a dramatic event finally forces them to do something. A good example of this is the willingness of a larger community to tolerate problem bars and taverns in central city areas while closing down those in suburbs. That results partly from homeowner complaints, but it also reflects unofficial acceptance of misbehavior based on misplaced stigmas.

In short, careless stigmas interfere with crime control by misleading us about who is the problem. Even though some people do more than their share of crime, most are in the vast middle, neither heavily criminal nor totally innocent. Tens of millions of Americans commit just a little crime, which adds up. So forget the stigmas or halos. Regard the people in your midst with a moderate dose of benign suspicion.

MAKING SENSE OF CRIMES
THAT SEEM IRRATIONAL

Even when an offender's motive has an abnormal twist, basic human tendencies show through. People differ greatly in their sexual tastes, but one person's orgasm is the same as another's. The fellow with unusual lust knows to pursue it when nobody is looking. Whatever the forbidden features of crime, the offender uses *practical techniques* to perform illegal tasks in the context of routine legal activities (see Rossmo, 2000). Often a criminal act will seem senseless to the outside observer. It is hard for that observer to accept that the offender really made a decision. But how can we build crime analysis if crime is irrational or fuzzy? We must find out how the offender thinks, even about bizarre violence, or we should find another line of work.

Violence, Too, Results From Decisions

Robbery is perhaps the easiest violent crime to explain: "Your money or your life" might not be polite, but at least it makes sense. Other types of violence are sometimes not so easy to explain. In their book *Violence, Aggression and Coercive Action* (1994), James Tedeschi and Richard Felson conclude that *all* violence is instrumental. They show how various types of violence have a purpose that responds to immediate situations. Tedeschi and Felson deny that *any* violence is "expressive." With a violent act, an offender can

- Get others to comply with his wishes
- Restore justice, as he perceives it
- Assert and protect his self-image

This is true even if the offender has only a split second to react. Suppose three 14-year-old boys throw a rock at a middle-aged man, cutting his arm. His neighbors conclude that "This is senseless violence. These kids did not get one cent for what they did." But the youths probably have a purpose. They may be punishing the man for yelling at them the other day. They may be putting on a show, proving how strong they are. Tedeschi and Felson (1994) show that various principles of social psychology can explain this and other violent incidents quite easily. Violence requires neither a unique theory nor an elaborate one and can be explained with the three instrumental explanations listed above. Unfortunately, many of the people who write about violence are so upset that they cannot calm themselves down enough to analyze it well.

What if two guys get into an argument? They get madder and madder, until one hits the other. Is this irrational violence? Regrettably, the words "rational" and "emotional" are often seen as opposites, implying that a person cannot make a decision while angry. Nothing could be further from the truth. Angry decisions are part of life. That does not mean an emotional person reasons carefully or for long. Nor are emotional decisions wise for the long run. But they are still decisions. (On this point, see Bouffard, Exum, & Paternoster, 2000; Harding, Morgan, Indermaur, Ferrante, & Blagg, 1998.)

For example, violence by athletes has a structure. They seldom attack the officials or the most muscular opponents. They almost always drop bats and sticks first. They punch, but don't kick. Athletes are well aware which acts of violence draw the worst punishments and stop short. They also pay attention to whether the officials are looking.

Crimes That Seem Bizarre

Kim Rossmo (2000) has shown that serial killers and rapists, no matter how bizarre they may seem, have a very practical modus operandi. They clearly make decisions, searching for targets near or on the way to home, work, school, or recreation.

What about the robber who kills someone for a single dollar? He risks life in prison or even the death penalty for nothing. Why? The offender usually does not know exactly where a criminal act will lead. Consider this example of a *blunder sequence*:

1. A robber accosts a man alone and demands money.

2. The victim gives him a dollar.

3. The offender gets mad, shoots the victim, grabs the wallet.

4. The offender runs away, only to find the wallet empty.

5. The victim dies from the gunshot wound.

An offender gets from here to there in steps and stages. Each act along the way made some sense from the offender's viewpoint, even if the endpoint made him look the fool. The community learns the endpoint but not the steps that led to it.

Consider an offender who stole a brightly painted yellow cab, then drove it 100 miles per hour, crashing it with loss of life and sustaining serious injury himself. How do you explain that? Imagine this scenario: Three guys are getting drunk, talking of their exploits. They brag about which women they have slept with. Charlie turns to talking about the cars he has stolen, and the others join him, going through various makes and models, each trying to top the other. Alan says, "I'll bet you never stole a taxi." Bill dares Alan to steal a taxi. Alan says he's not afraid to do so. The slope gets even more slippery as Charlie calls him on it. "Let's go look for a taxi now." After they find a taxi and steal it, Bill says, "See how fast it will go." Each decision has a certain logic, such as looking important or challenging the other guy. None of the three foresaw where it would all end.

Maturity brings with it wisdom and foresight, but don't let that get in the way of the crime analysis. To understand illegal actions, you must break down the steps an offender takes, no matter how immature or unwise, as each decision gets a person into deeper trouble.

Puzzling Sexual Offenses

It seems entirely bizarre and irrational that some offenders molest young boys, have a fetish for women's shoes, seek sexual intercourse with dead bodies, or engage in serial rape. Yet these offenders, however unusual, prove to be quite practical in carrying out and concealing their crimes (e.g., see Mannon, 1997). But how could a person arrive at these "strange goals" through a routine process? For example, why would a rapist attack a woman's shoes rather than her body?

To understand how such goals develop, bear in mind that sexual behavior can be improvised, incremental, and secret. Most men begin their sexual lives influenced by legitimate stimuli (such as a woman's perfume or shoes as a sexual *reminder*), then transfer their interests to women themselves. But a few men fail to make this transition, only to be left with a shoe fetish. Reinforced by masturbation, some individuals drift farther and farther away from approved urges. Even the most unusual or extreme fetish or macabre behavior can develop secretly and incrementally through routine steps, powerfully enhanced by sexual rewards. Once more you see that emotions and reasoning go together.

The fact that offenders make decisions and do so with practical matters in mind also applies to more conventional offenses. But can society keep people following the rules by assigning them to conventional and well-approved activities and roles?

SOCIAL ROLES, TIES, AND CRIME

Weddings are heartwarming. Bride and groom are full of beauty and innocence, with a perfect future ahead. It is easy to tell yourself, "I wish George would settle down and get married. The right woman will get him to stop drinking and keep him out of trouble." In the next section, we discuss whether these "wholesome roles" really reduce crime, and if they do so, how.

Wholesome Versus Offending Roles

Among the traditional wholesome roles are marriage, work, and joining the army. But when it comes to preventing crime, a lot can go wrong. A man can marry a female alcoholic or offender. He can abandon a good wife for the bottle, his nasty friends, or both. He can beat his wife, just as he beat his girl-friends. He can join a young and wild army outfit. (Many men came back from Vietnam with hard-drug habits.) He can be in the right places for avoiding crime, but still make the wrong decisions. Chapter 6 tells us how jobs for young people often backfire, producing more criminal behavior rather than less. School enrollment can also produce many crime opportunities. Chapter 7 shows us how work roles also can backfire for adults, producing greater opportunity to commit various crimes.

The essential problem is that roles we think of as wholesome do not always confront people with fewer crime opportunities and, indeed, sometimes offer more. Because we feel that a wholesome role ought to make people good, it is easy to forget that it might not do so at all. Only by specifying the details can you understand how marriage and other such roles could indeed reduce crime.

When Family Helps the Most

Despite the widespread attention given to family violence, strangers tend to be much more dangerous, considering the little time one spends among them. Time budget studies tell us that people devote on average about 1 hour a day with strangers, compared to 14 hours per day with their relatives. That single hour with strangers generates, for males, 50 times as much assault risk as an hour spent with relatives. For females, an hour with strangers is more than 30 times as risky. Don't be taken in by the family violence industry: You are safer at home. Homeless people have proven this for us by sleeping outdoors where strangers can get at them (Hagan & McCarthy, 1998; McCarthy, Felmlee, & Hagan, 2004; McCarthy & Hagan, 2001, 2005; McCarthy, Hagan, & Martin, 2002); their victimization rates are astronomical.

When a young person moves away from home to get an apartment, that turning point produces a vastly greater risk of victimization. Changing family roles can provide crime turning points not by improving character but by altering everyday supervision of people and property. Such supervision can be measured. Asking how many nights a week young people go out is a simplistic start, but better than nothing. A much more precise research effort is reported in Felson and Gottfredson (1984), where people were asked about their teenage years, including what they were doing, where, when, and with whom (see Chapter 6).

Marriage vows have moral, emotional, and religious significance. But marriage is also a living arrangement. That's really what's important for crime. Even if not always so dependable for lowering offending, marriage clearly reduces crime *victimization*. This was first discovered in the late 1970s and has been confirmed in one victim survey after another. Two adults living together as a couple, including those who are married, have a physical advantage in thwarting crime. They are significantly less likely to be personal targets of crime, suffer property crime, or have their cars stolen. They can more readily watch the area near their home. They are more likely to go out together, and hence are not as easily picked off by offenders. Their homes are less likely to be empty when burglars check. Married people spend more time in safer settings, such as the home, in contrast to unmarried people, who spend more time socializing in nonfamily, nonhousehold settings. If married people get drunk, they probably do so at home where they are *relatively* much safer per hour spent intoxicated. Even when married people leave home, they are more likely to go to safer settings.

Keeping Bad Company

Going around with the "wrong crowd" is often associated with criminal behavior and juvenile delinquency. Do innocent boys learn their delinquency from bad boys? If so, how did the first boy get into trouble? Does each delinquent produce 1 other, 3 others, or 10 others? The simple correlation is well established between being delinquent and having delinquent friends. It's harder to explain why.

Travis Hirschi (1969) suggests that delinquents pick each other as friends afterwards, that is, "birds of a feather flock together." Exhibit 3.2 offers seven specific reasons why an offender might want co-offenders around to help carry out their illegal tasks (Tremblay, 1993; Warr, 2001, 2002). Two can intimidate better than one (the same role). Two can perform different roles, one kicking in doors while the other serves as lookout. Offenders can diffuse blame or help each other learn about more targets. Group effects also feed their delinquency. Offenders provide audiences for one another and egg one another on. Delinquent friends are like intoxicants, their presence serving as a disinhibition against misbehavior. In general, delinquent friends serve a purpose for one another.

Exhibit 3.2 Why One Offender Might Have Others Around

Reasons	Comment
1. To perform same role in crime	Two intimidate better than one
2. To perform different role in crime	Joe kicks in door, Fred is lookout
3. To diffuse responsibility	Shared blame is evaded blame
4. To learn about more targets	Two pairs of eyes better than one
5. Mutual bad influence	Well behaved alone, in trouble together
6. Audience for prowess	Many offenses are used to show off
7. Breaking rules more fun together	Getting high is less fun alone

CONCLUSION

You can see that settings change choices by providing temptations and controls. These are mediated by various tangible cues that tell people what they might get away with. People make decisions accordingly. They are somewhat constrained, but not necessarily all day or all week. The constraints shift from one setting to another. By making your insights about crime control more tangible, you are better able to examine whether and when people will make illegal choices. That enables you to avoid the vague or false hypotheses about how wholesome roles prevent crime, while understanding better how marriage reduces victimization and delinquent friends assist one another in illegal action.

We have tried to make our understandings of crime decisions and settings more explicit. But these decisions must not be taken in isolation. They are indeed part of a larger system of community life, as the next chapter will show you.

MAIN POINTS

1. Offenders make decisions seeking to gain pleasure and avoid pain. Even so, they respond to the settings that limit their choices. An offender is neither too careful nor totally spontaneous or without reasoning.

2. Most offenses pay in the very short term, namely, on the day they are committed. In the long run, most offenders suffer enough bad experiences to justify our conclusion that crime eventually fails them.

3. Offender freedom to decide is greater at some moments but less at others. Settings offer different choices at different times. Offender choices can shift within a particular setting, in that one bad decision can lead to others down the pike (see Chapter 8).

4. Environmental cues are needed to assist self-control. Most of us need reminders to follow the rules. A reminder is nothing more than a cue designed for people who have a good deal of self-control and are basically inclined to follow the law.

5. Stigmas are a poor substitute for environmental cues. People use careless stigmas based on looks, guesses, skin color, or sloppy rumors that can interfere with crime control by misleading us about who is the problem.

6. Criminal acts often seem senseless or even bizarre, but offenders still make decisions to commit them. For example, violence is not irrational but has goals and quick calculations, even when the offender is very angry.

7. Traditional, wholesome social roles and ties such as marriage, work, and joining the army may not always discourage, and may even encourage, criminal decisions.

8. Considering the time spent with family, individuals are much less likely to be victims of crime in a familial setting.

9. Delinquents often associate with one another, but it is not always clear why. One might influence the other to be bad, or they might choose one another because of their common delinquency. In addition, they might assist one another in delinquency or have mutual negative influences.

PROJECTS AND CHALLENGES

Interview projects. (a) Interview anyone with experience in a sport with some amount of violence. Find out the decision factors and settings that invite violence, as well as those that discourage it. (b) Interview a fellow student about settings inviting or discouraging illegal action. Do not ask questions about personal involvement that would put that student at risk. (c) Interview a bartender, former soldier, or high school teacher about techniques for making the peace.

Media project. Find a media account of two conflicts in baseball, one that escalated and one that did not. Consider decisions the baseball players or umpires might have made and how the escalation might have been avoided.

Map project. Sketch a map of a convenience store from above. Put into the sketch five features that might influence someone's decision to rob it or not. Number these items in order of likely importance to the thief. Justify your numbering.

Photo project. Photograph three settings that are safe for families and three settings that are not. Discuss why.

Web project. Search the Web for sites that discuss conflict. How many of them treat conflict as irrational and how many, in some sense, as rational and/or structured?

NOTE

1. Brother of the senior author [MF], and author of the world's best research on violence.

FOUR

BRINGING CRIME TO YOU

———•◆•———

Many college students are aware that they live in a different world from the rest of the country. A large college or university campus can reflect and blend the history of daily life. Not all campuses are like this, but some are:

1. Like a village, college students are often on foot and are in frequent contact with others they recognize, living in apartments or dorms divided into small units.

2. Like a town, bicycles extend the range of daily travel and activity and increase the number of people who can converge on a given spot.

3. Like a city, buses integrate the campus with a much larger local population.

4. Like a suburb, autos extend still farther the daily commute of students and staff, linking campus to metropolis.

If your campus has these traits, you are experiencing the history of human settlement covered in this chapter. As we present that history to you, bear in mind that this history may be alive in your life today and the crime problems you have to face. If your campus is *not* like this, at least try to imagine the four stages described below and what they mean for crime.

Recall our discussion in the previous chapter about the cues in everyday life that invite or discourage crime. These cues emerge from the forms of daily life. Crime feeds off the *physical* form of local life, whether in a village, town, city, suburb, or university campus. That form is organized by how people and things move about every day in systematic ways. As people move through time and space, they come into contact with products, some of which are stolen (see Chapter 5). They encounter cues that stimulate temptation and/or enhance control. Everyday life thus sets the stage for people to break laws, hurt each other, and even hurt themselves.

FOUR STAGES IN THE HISTORY OF EVERYDAY LIFE

The four stages in the history of everyday life stem from the transportation technology of each era (see Felson, 1998; Hawley, 1971). These stages help us understand the growth of crime and the forms it takes.

1. The Village

Most people traveled on foot, and their daily range of activity was less than 4 miles. With most villages less than about 250 people, daily interaction was entirely local and people knew each other. Strangers could seldom assemble in large numbers. What few things people owned were custom-made and easily recognized if stolen. Local crime was unlikely, but villagers suffered from marauding bandits and highway robbers.

2. The Town

When horses were domesticated, people could travel about 8 miles a day. Local populations could exceed 10,000. Most townspeople would still recognize one another by name or have a friend in common. Local crime was limited, but horses made speedy raids possible. Horses and wagons themselves became targets for crime. Overall, towns provided considerable security.

3. The Convergent City

After nautical technology advanced, ships transported voluminous and valuable goods to leading port cities. Their docks and warehouses provided important crime targets and fed a tremendous crime wave (Colquhoun, 1795/1969). These cities grew greatly in size, even then spawning such dangerous neighborhoods as the rookeries of London (Brantingham & Brantingham, 1984). But that was just the beginning. Railroads, powered by fossil fuels, fed even more people and goods into even more cities. Elevators sent buildings upward, while steam-powered factories—operating around the clock—concentrated workforces. Compared to village and town stages, the convergent city drew far more strangers into mutual contact, with greater risk of property and violent crime.

4. The Divergent Metropolis

Automobiles allowed people to travel farther and wider than ever before. No longer limited to train tracks, they filled in the areas in between. The ability to supervise space and property declined. Autos themselves provided vast crime opportunities—theft of cars, parts, and contents; use of cars to carry out felonies; passenger exposure to personal victimization; crimes in streets and parking areas; and more. Later, we will discuss some modern adaptations of these four stages of urban history, going beyond the divergent metropolis. But first, let's consider the convergent city further.

LIFE AND CRIME IN THE CONVERGENT CITY

The convergent city provided not only crime opportunities, but also sources of control.

Safe Aspects of the Convergent City

The convergent city might have seemed risky compared to the village and town stages of the past. But its crowds of pedestrians also provided natural surveillance of streets, discouraging overt physical attacks most of the time. Street vendors, including newsstands, enhanced this surveillance (see Chapter 9 for further discussion of people on the street who enhance natural surveillance).

Initially, convergent cities were slow to transport people, with horses and buggies producing terrible congestion and ongoing concerns about where to step. However, this changed as public transit became mechanized. Subways, streetcars, commuter trains, buses, and taxis moved people around in the convergent city. These vehicles funneled crowds into work, school, shopping, and entertainment districts. The vehicles carried limited crime risk within, even if people might have sometimes felt more at risk there because of their contact with large numbers of total strangers.

As public transit systems developed in convergent cities, they filled up very little public space considering the number of people they moved about. A single urban train line could carry as many commuters as a six-lane highway of today. Subway cars took up much less parking space than automobiles. Thus, convergent cities had a chance to design and control space to minimize urban crime.

Risky Aspects of the Convergent City

Despite the fact that pedestrian crowds in the convergent city would discourage some crime, they still posed threats through exposure to strangers (Riedel, 1999). Pedestrian crowds make it easy to pick pockets and slip wallets out of purses. Grab-and-run attacks on merchants' wares were also common. Crime in general is a stop-and-go activity (Felson, 1998), and convergent cities helped a thief get back into the flow of foot traffic for a quick exit. Public transport could also assist the offender in a hasty departure.

Even if train cars and buses were relatively safe inside, they generated crime in important ways. Their stations and bus stops were not always safe. Long subway platforms, built to handle the worst crowds, also assisted offenders by making surveillance difficult. Old systems had many nooks, blind stairwells, and shadowy places, making the offender's task easy (LaVigne, 1996). It is fairly safe right by the station, which has enough foot traffic to be safe. But areas a block or two away from the station have too few people around for constant natural surveillance and are the most dangerous: Offenders can wait there for stragglers (Block & Block, 2000; Smith, 2005; Smith & Cornish, 2006). Furthermore, side streets and alleys are not frequented by crowds. After the Industrial Revolution, many cities had late shifts of factory workers; that meant that offenders could find the right time and place for attack.

The convergent city afforded settings for public drinking, often in crowded taverns with high risks of bumping and fighting inside. Taverns could dump crowds or individuals onto the street in a drunken stupor. Those drinking in streets or on stoops could create problems for themselves and others.

Initially, alcohol problems were greater in rural areas, where the grains were grown. But better roads brought alcohol into the city, which was least able to manage it because of the crowds.

Convergent cities provided excellent opportunities to mismanage parks, street corners, and other public places. With drunken young males hanging out, crowds could turn from a source of supervision to a font of trouble. Even three or four drunks together could be rather hard to handle. At the same time, the entertainment districts in cities of the past had street crowds combining people of various ages, not all drunk.

You can see now that convergent cities provided sources of both security and crime. All of life is trial and error. In every stage of life, people seek to prosper and to keep others from seizing their gains. Couples want to raise a beautiful daughter, but they also want most men to leave her alone. Urban life offers prosperity, unfortunately with a lot of insecurity. But people keep trying and sometimes find solutions.

The Urban Village—A Low-Crime Area

In 1961, Jane Jacobs published a classic book, *Death and Life of Great American Cities.* Her thesis was that the old urban neighborhoods, despite a bit of grime, were actually good places to live and raise children. She was adamantly opposed to the urban renewal projects that bulldozed these neighborhoods and replaced them with high-rise public housing. She felt that the old neighborhoods were built for pedestrians and that life on the street created not only a vibrancy of a living city but also a relatively low crime rate. She anticipated that high-rise buildings and streets built for cars but hostile to pedestrians would destroy neighborhood life and ultimately undermine the city as we knew it.

About the same time, Herbert Gans (1962) invented the term *urban village* to describe areas within the convergent city where people could find rather secure daily life and face-to-face interaction. Combining the ideas and observations of Jacobs and Gans, an urban village provided

- Stable ethnicity
- Stable residency
- New entrants recognized from the past
- Homeownership, even if homes were tiny
- Narrow streets and wide sidewalks

In the urban village, middle-aged adults controlled urban spaces, not the toughest youths. Women had more power than met the eye, keeping the men largely out of trouble. People knew each other and could not get away with much. Owners and long-time renters together watched over the area (note the discussion of place managers in Chapter 2). New entrants had social ties to those already there. Streets were well supervised. The paint may have been shabby sometimes, but the streets were clean. And you did not have to be rich or even middle class to be safe.

To understand why stable residence is important, take a look at Exhibit 4.1, which compares short-time residents to those living at the current address 5 years or more as of 2005. Short-time residents suffered 2.2 times as much burglary, 2.4 times as much vehicle theft, 2.1 times as much general theft, and 3.6 times as much violence (see Felson, 1998). Stability and home ownership, combined with the other factors mentioned above, made urban villages safer than the rest of the convergent city.

Exhibit 4.1   Property Crime Victimization by Time at Current Residence, United States, 2005*

	Time at Current Residence	
Type of Crime	Less Than 6 Months	5 Years or More
Household burglary	57.9	26.2
Motor vehicle theft	17.1	6.7
General theft	221.9	103.8

SOURCE: Bureau of Justice Statistics. (2006). *Crime Victimization in the United States, 2005* (Statistical Table 51). Washington, DC: Author. Retrieved August 14, 2009, from http://www .ojp.usdoj.gov/bjs/pub/pdf/cvus/previous/cvus51.pdf

*Property victimization rate per 100,000 households.

Whatever our nostalgia for it, the convergent city produced congestion, noise, annoyance, and some types of crime. Its high density ensured that you could hear ambulances and police sirens day and night and that other people's problems would impinge on you. A visit to Manhattan is all you need for a quick reminder that a city can get on your nerves. Although the larger city offered chances for boy to meet girl, including ongoing entertainment districts, young singles were too poor to have their own apartments and had to go home to Mother. Boyfriend and girlfriend could not go home for the night. More settings were then under control of middle-aged people, especially women. Even after marriage, most couples had to live at home for at least a while. For very tangible reasons, crime rates were relatively lower in convergent cities of the past than in the metropolis as we know it today.

Prosperity and growth changed that world. The twin searches for affordable real estate and peace of mind help explain what followed.

CRIME AND THE DIVERGENT METROPOLIS

A great transition has occurred in the past 50 years: The convergent city gave way to the divergent metropolis. This transformation was set in motion by events that occurred before the transition began. Many people contributed, but three noteworthy events help explain best what happened.

Three Inventions With Real Consequences for Crime

Although social change is always complicated, three important inventors helped produce the divergent metropolis and life in America as we now know it.

- Henry Ford invented the modern assembly line. Producing cars by the millions, Ford made them available to the masses.
- Serving in the navy during World War II, William Levitt learned how to put up wartime buildings quickly. After the war, he applied these lessons to invent mass-produced suburban housing. In building Levittown on a former potato field in Long Island, New York, Levitt organized the opposite of a moving assembly line, with specialized workers moving from one site to the next.
- In 1919, Dwight D. Eisenhower was a 29-year-old officer for an army truck convoy crossing the continent. The trip was so slow and inefficient that he never forgot it. Thirty-seven years later, as president of the United States, Eisenhower signed a bill that inaugurated the interstate highway system. On that basis, we can designate him the most influential peacetime president of the 20th century.

The auto industry built so many cars, the home industry so many suburbs, and the highway system so much access that daily life in modern America was transformed. These three innovations also produced the major crime opportunities of today. Cars and suburban homes provided millions of new targets for crime and conveyances for offenders. Highways provided crime access and evasion of control at a level never before seen.

It Did Not All Happen at Once

With housing and roads, the metropolis could and did develop. First, bedroom suburbs grew along travel corridors into the central city. But these suburbs began with low crime rates. Youths did not yet have autos, suburbs had not grown together, mothers were outside the labor force, and shopping malls had not developed. With the dispersion of jobs and shopping to the suburbs, and the other conditions reversing, suburbs became much more suitable for criminal activity. In time, suburbs had more to steal, more people to steal it, and easier transport for those most likely to do so. People got home later and had less time to watch over youth or property.

The divergent metropolis developed through an intricate unpacking process, carried out over many years, that involved

- Dispersing homes and other buildings over larger lots, wider roads, and vast parking areas, all at lower heights. These were increasingly difficult to protect from criminal entry.
- Dispersing people over a greater number of households, with fewer persons living in and supervising each.[1]
- Spreading people over more vehicles as they travel and park. That subjected parked vehicles and drivers walking to and from parking spots to high risk of attack.
- Disseminating social and shopping activities away from home and immediate neighborhood. As strangers, people became ready targets of crime or perpetrators.
- Spreading vast quantities of retail goods over a wider expanse of space, with fewer employees to watch them, thus inviting people of all ages to shoplift.
- Assembling millions of dollars worth of cars in huge parking lots with virtually nobody to watch them.

Ironically, we think of cars as private, yet they require much more public space than does public transit. That space is then difficult to control and thus invites crime to occur at higher rates. More generally, the divergent metropolis brought offenders and targets closer together in locations unsuitable for supervision.

REAL LIFE OUTGROWS FOUR STAGES

Life means growth. And growth tampers with any categories we give it. Not only have we seen urban villages, but also new forms of old cities. The holdover city of today combines features of the convergent cities of the past with the divergent metropolis.

The Old-But-New City

Today's old American cities are a *remnant* of their convergent past, with subways and buses still moving people about. Old buildings are still tall, and new buildings may be even taller. But today, these cities also have cars rushing about their narrow streets and people trying to find parking spots. Suburbanites drive into the city if they can afford parking, or drive to train stations and enter that way. Nowhere is this more evident than in New York City—a hybrid of past and present.

Most European capital cities—Paris, London, Amsterdam—are also hybrids. European urbanites tend to own cars and drive them daily, especially in cities that are not capitals and have few American tourists forming mistaken impressions. Europeans have a love affair with the automobile. They just have less room to spread out than we do. Throughout the world, old cities incorporate cars and related problems and benefits. Hybrid cities seem to have more crime problems than convergent cities of the past. Their pedestrians can find plenty of cars to break into. They have enough people on the street to get mugged, but not as many to act as guardians against crime.

The Center Harmed by the Suburbs

After the divergent metropolis has grown, the remaining central city is just a shell of its own past. It has lost its wealthy residents and even its middle class. Its tax base is decimated—very relevant in the United States, where local taxing authority is extremely important. Jobs have fled to the suburbs. Shopping malls have all but destroyed urban shopping. Even the city's own residents go out to the suburbs to shop as much as they can. Who can blame them? The divergent metropolis has sucked life out of the central city.

The Great Metropolitan Reef

The earlier development of the divergent metropolis followed main road arteries and formed a patchwork of development. The metropolis grew like an amoeba, surrounding and absorbing each nearby town, allowing people to live at suburban densities while using automobiles to make up the distance. In time, the areas between the lines and patches filled in with additional development: residential, shopping, factory, and office. Growth continued outward, and towns grew together. In time, even metropolises themselves begin to fuse into a single organism, a seemingly endless suburban sprawl at moderate- or low-metropolitan density.

Thus, the divergent metropolis becomes part of a Great Metropolitan Reef. Each new piece of suburb fastens onto this *metroreef* like coral, building outward and connecting inward. The metroreef not only grows like coral but also functions like a coral reef after growth has taken place, with organisms moving about the reef continuously and diffusely. The same freeways that allowed quick access for shopping, work, and friendship also made many types of crime easier, including burglary, fencing stolen goods, or assembling a group of drunken youths. The suburbs were no longer a refuge from urban problems. Malls provided many crime opportunities formerly associated with the central business district of the convergent city. Indeed, metropolitan specialists have begun to use the term *edge cities* to refer to growth of commercial districts outside the center. You can see once more that our categories can barely keep pace with human inventiveness.

Convergent Cities Coming Back in a New Form

In the past few years, we have seen a resurgence of many core cities. Baltimore built a new National Aquarium and a wonderful new baseball stadium in the old-time style. Excellent urban renewal has made the center attractive, even if the area around it has much work to do. Cleveland, too, has made important strides toward recovery. Grogan and Proscio (2001) write of these "comeback cities," and we have reason to believe more will follow. But most of the metropolitan population is still scattered away from the original center. We are more likely to see an urbanization of suburbs than a rebirth of the core.

Many cities are enjoying major declines in crime rates, but it remains possible that their new commerce will bring with it additional crime targets and higher crime rates in relation to their populations. New urban forms are in progress, and they could lead to additional crime opportunities.

POPULATION DENSITY, SHIFTS, AND PATTERNS OF CRIME

Patterns of crime are not so easy to study in a metropolis, where people and things move around so much. Once it was believed that crime rates vary directly with population density. In other words, criminologists expected to find more crime in high-density cities than in low-density cities. But it is not so simple. Decker, Shichor, and O'Brien (1982, Table 4.1) found fantastic differences in how population density related to each type of crime (see Exhibit 4.2). Cities with high population density tend to have much less burglary and household larceny than cities with low density. Motor vehicle theft, however, goes up somewhat with population density. Robbery and larceny with contact (often similar to it) go up a good deal with a city's population density. Here's the puzzle: Why does density correlate +.57 with contact-larceny but −.77 with household larceny? It is puzzling to see these signs going in opposite directions. To solve the puzzle, consider the modus operandi of each type of crime.

Why Does High Density Reduce Residential Burglary?

Household burglars have favorite ways to enter homes. They choose rear doors or windows two thirds of the time, according to a study of 1,284 burglaries by Winchester and Jackson (1982). Exhibit 4.3 goes on to show risks for different types of housing. Detached houses have almost three times the burglary risk as flats (apartments) and eight times the risk as long terrace apartments with no available rear or side entry. Clearly, burglars pick housing types carefully enough to escape the eyes of neighbors. Importantly,

Exhibit 4.2   Population Density Correlated With Crimes

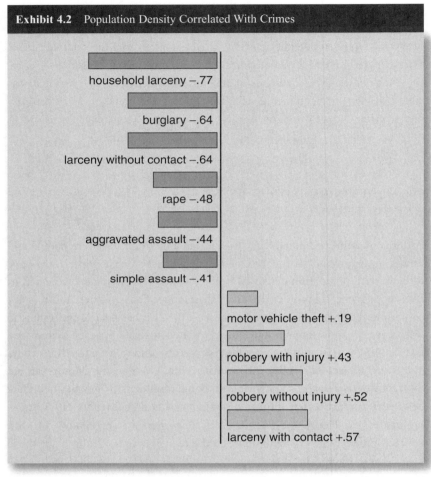

SOURCE: Adapted from Decker, D., Shichor, D., & O'Brien, R. (1982). *Urban Structure and Victimization* (Table 4.1). Lexington, MA: Lexington Books.

high-density residences are the hardest to break into, and those in low-density areas are the easiest targets. The detached houses of low-density cities and suburbs offer

- Entries on all sides,
- More space between, and
- Trees and bushes to block the view of doors and windows.

Exhibit 4.3 Burglary Risk for Different Types of Housing

Type of housing	Implied density	Burglary risk per 1,000 homes
detached house	low	
bungalow farmhouse	low	
flats	medium	
short terrace apartments	high	
long terrace apartments	high	

SOURCE: Calculated from data from Winchester, S., & Jackson, H. (1982). *Residential Burglary: The Limits of Prevention.* London: Home Office.

The lower the suburban density, the more windows, space between homes, and bushes to block sight lines. These homes are secure and comfortable for burglars.

On the Other Hand, There's Vehicle Theft

Why, then, does motor vehicle theft go up a bit with population density (i.e., they have a moderate positive correlation)? Although there are fewer automobiles per capita in dense cities, they are parked in more risky places. Drawing from Clarke and Mayhew (1998), Exhibit 4.4 shows the risk of auto theft per million hours parked in a particular type of spot. Generally, a car is safest from theft when you are at home, less safe when you are working, and least safe when you are doing something else. If you park somewhere to shop or for leisure activities, watch out! Indeed, if you park in a public lot for these purposes, your risk is almost 190 times as great as parking in a personal garage.

Exhibit 4.4   Auto Theft Risk in Different Places People Park

Activity when parked	Parking location	Thefts per million hours that cars are parked there
residential	personal garage	◖
residential	apartment garage	◖
residential	home carport, drive	◖
work	work garage	●
residential	apartment lot	●●◖
residential	home street	●●◖
work	work street	●●◖
other	street	●●●●●●◖
other	public lot	●●●●●●●●◖

SOURCE: Calculated from data from Clarke, R. V., & Mayhew, P. (1998). Preventing crime in parking lots: What we know and what we need to know. In M. Felson & R. B. Peiser (Eds.), *Reducing Crime Through Real Estate Development and Management.* Washington, DC: Urban Land Institute.

NOTE: Each dot represents one car.

In many European cities, as well as dense American central cities, people have cars but no private garage, carport, or driveway. Giving up a parking spot means having to park farther away from your apartment and increasing risk that someone will steal your car or its contents. In contrast, low-density areas provide personal garages and opportunities to park right near the home or office. You can now understand why rising population density brings *less* burglary but *more* auto theft.

Apartment garages and carports are not as good as a personal garage or as bad as open apartment parking lots or streets near home. The more enclosed and private your parking, the more secure it is. Your own personal garage is by far the best of all. Many people have used garages for storage and parked in the driveway outside. This is a mistake if you want to reduce auto theft risk.

Other Features of Density

It also is not difficult to see why personal larceny and robbery go up with density: Urban crowds make it easier to pick pockets and grab purses. In addition, high density fosters small groceries, which are easy for robbers to knock off during times when patronage is light. With more people on foot (and hence more stragglers at dusk and after dark), you can also expect more personal robberies in high-density cities.

Note that density affects robbery differently from assault, even though both are violent offenses. An assault backs the victim into a corner and can easily generate screams or shouts, which are likely to be heard in a high-density area. The assailant benefits from a wider expanse of empty space in a less dense area. A rapist certainly needs some time and privacy, even if acting outside. Although a robbery also requires a realm of isolation, a shorter time suffices. Waving a gun in the face of a clerk or straggler gets the money, no questions asked. Thus, high density is less likely to impair robbery than assault.

Density Shifts Quickly

In a modern society, population density shifts very quickly from one spot to another. An office area packed with people on weekdays is abandoned later in the evening and on weekends. A shopping area largely empty during school hours is dominated by teenagers after school, and then changes when adults come home from work. As these populations shift, so does crime. With population density shifts come related shifts in crime opportunities.

All crime is local. The minimal elements of crime converge locally. The physical and social components of crime are fluid in the course of a day, a week, and a month, posing quite a challenge to crime analysis.

Density variations are so great in modern life as to dangle very different crime temptations. Manhattan today has about 70,000 people per square mile—many more during daytime hours. Millions have moved from New York City to live in New Jersey suburbs or others surrounding the city. New Jersey is America's most urbanized state, at more than 1,000 people per square mile in the state at large. Its local population shifts quickly, as people leave to work in New York City, Philadelphia, or parts of New Jersey other than the place they reside. When they go to work, burglars check out their homes, and near work, other offenders look over their cars.

CONCENTRATED ADVANTAGE
FOR COMMITTING CRIME

Exhibit 4.2 compares population density and crime among different cities. But within any given metropolitan area, denser central city areas tend to have more crime, especially violence, than the suburban rings around them. Yet its youths are no more likely to report committing offenses in standard surveys! This paradox is resolved as follows: The central city helps some offenders to start at younger ages, stay more years in active offending, and get deeper into trouble. A subset of youths does most of the damage. In selected localities, that subset may not be so small, but it seems hardly fair to pick the worst individuals to represent their neighbors.

Shaw and McKay Gave Us a Start

On the other hand, the worst have a lot to teach us. They provide a natural experiment: How can society concentrate crime opportunities to maximize misbehavior? Burgess (1916) and White (1932) were well aware that local areas could greatly enhance delinquency.

Shaw and McKay (1942) were most famous for showing this in their studies of "delinquency areas." They noticed in particular that these areas had many abandoned buildings, declining industries, and out-migration of people who were getting somewhere, along with in-migration of new groups. Each new group developed high crime rates as it entered the "worst" part of town. Shaw and McKay were not always precise when writing about "social disorganization," but they gave us an important starting point for precision by specifying that abandoned buildings assisted criminal behavior. We know much more today to help understand how central city areas deliver extra crime opportunities.

The Lack of Place Managers Gives Crime an Advantage

Recall that Chapter 2 discussed the significance of place managers for preventing crime. Earlier in this chapter, we considered how urban villages provided place managers, namely, more homeowners and long-time residents. But on the other side of the tracks, there are few homeowners or long-time residents to watch over people, places, and things. Apartment buildings cannot afford doormen. Many public housing projects are found here; although some may be well designed and managed, many are not (see Chapter 9). Park and playground supervisors have long been cut out of city budgets, with parks too often taken over by thugs.

Industrial Land Use Assists Crime

Mixing residence with industry generally makes it more difficult to keep young people away from delinquency (Burgess, 1916; White, 1932). That's even truer when central city industries have gone bankrupt or found greener pastures, leaving a shell of abandoned industrial sites ideally suited for crime to occur. Prostitutes use these sites for strolls and assignations; drugs are sold and used in and near them; boys hang out, get drunk, or fight there; and remaining items are stolen and sold, feeding substance abuse. Industrial properties also provide escape routes to offenders. That makes this turf virtually impossible to police.

One of the main problems for inner-city police in old industrial areas is the inability to corner and arrest an offender who has 8 or 10 escape routes. Similarly, customers for illegal goods find convenient entry and exit, also with little risk of successful arrest. Other industrial areas provide parking lots for local youths to plunder during daytime hours, with good spots to chop up cars at night.

Exposure to Skidders

Skidders are people who go downhill in life, perhaps because of drugs or alcohol. But they may also have suffered a variety of diseases, mental illness, accidents, war injuries, or anything else that interferes with their prospects in life. Skidders very likely end up poor and may also find themselves easy for offenders to pick off. Some may become offenders themselves, but leave that aside. They contribute to higher crime rates in poor neighborhoods simply by providing local offenders with excellent crime targets.

Sometimes we forget skidders when analyzing how poverty relates to crime. Many people who are poor today were not poor in origin. Thus, middle-class people or their children can end up on the other side of the tracks if they become alcoholics or drug abusers, or otherwise fail to succeed by traditional standards. Homeless runaway youths may come from middle-class homes. Indeed, high rates of crime participation and victimization are quite relevant to crime. As we study more, we learn more about *exactly how* poverty and crime can be linked.

More Cash, Less Credit, and Things to Steal

Poor people have poor credit. That means they carry more cash relative to their income than middle-class people. Poor people also spend a smaller share of their income on investments, home equity, real estate, heavy televisions, and many other goods that are hard to steal. With cash in their pockets and lightweight electronics as their best luxuries, they provide more adequate crime targets than you might guess from their income.

In short, society delivers a lot of crime to low-income areas. As Chapter 5 shows, too, that delivery does not depend on any extra crime tendency among poor people.

CONCLUSION

The divergent metropolis serves to unpack human activities. It disperses people over more households, more metropolitan space, more vehicles, and more distant activities. In the village and town stages, it was easier to keep crime low. The convergent city provided new crime opportunities but confined these to certain main locations; it also devised the urban village to contain crime. In contrast, the divergent metropolis weakened the supervision of space and provided even more crime opportunities. The vast increment in purely public space within the divergent metropolis undermined local people seeking a more secure environment.

Society is a living and learning organism. Each of these four stages—the village, the town, the convergent city, and the divergent metropolis—blends into its successor. Villages are absorbed into towns and towns into cities. The convergent city and divergent metropolis blend together. Living systems are in flux, and so are the crime opportunities they generate and the ones they reduce. By understanding these systems and how they grow and transform, you can better understand crime as a large-scale system for the very mundane and small-scale incidents that you think of as crimes.

MAIN POINTS

1. The history of everyday life can be summarized in four stages, responding to the transportation technology of the time. They are the village (on foot); the town (horses); the convergent city (ships, railroads, etc.); and the divergent metropolis (automobiles).

2. Transportation technology is the driving force of change in everyday life. As transportation modernizes, more people are brought into daily contact. That contributes to rising crime rates in many ways.

3. Even though the convergent city has more crime than do towns, its urban villages combat crime with more personal contact and stability. Crowds on the street often produce safety, at least when sober. On the other hand, convergent cities expose individuals to strangers, provide unsafe physical environments, and facilitate offenders' quick escape.

4. The "urban village" is an area within the convergent city where people can find rather secure daily life and face-to-face interaction.

5. Three important changes helped produce the divergent metropolis and life in America as we now know it. The auto industry built so many cars, the home industry so many suburbs, and the highway system so much access that daily life in modern America was transformed.

6. The divergent metropolis, with its reliance on cars, generates more crime than the convergent city. Cars not only make good crime targets but also lengthen trips and lessen control over targets and offenders alike.

7. The great metropolitan reef is the development of a seemingly endless suburban sprawl at moderate- or low-metropolitan density.

8. Density shifts and their impact on crime vary. Auto theft and burglary depend on very different densities and settings.

9. Lower income areas often draw more crime. The lack of place managers is more apparent in lower income areas where there are few homeowners or long-time residents to watch over people, places, and things. In these areas, industrial settings assist crime especially when they have been abandoned. Skidders contribute to higher crime rates in poor neighborhoods simply by providing local offenders with excellent crime targets. People who are poor have more cash in their pockets and lightweight electronics as their best luxuries, thus providing more adequate crime targets than you might guess from their income.

PROJECTS AND CHALLENGES

Interview project. Interview an elderly person about what village, town, or city life was like in the past. Ask especially about such topics as street life, people knowing neighbors, and activities of women.

Media projects. (a) Pick out five detailed media accounts of specific offenses for which nobody was caught. Avoid outlandish or unusual examples if possible. How did each offender probably take advantage of the surrounding area and setting? (b) Obtain a literary description of life in a large city of the past, one without automobiles and relying on streetcars or other old modes of transportation. Do a thought experiment about how theft would occur in such a city and the forms it would take.

Map project. Using Google Maps or Google Earth, access a satellite photo of a local parking lot and the building(s) it leads to. Put one asterisk (*) at the spots where the car itself is most likely to be stolen. Put two asterisks (**) where the car's contents are more likely to be stolen.

Photo project. Use the Street View in Google Maps to capture images of some buildings from past stages of history. What stage produced them? What tells you that? What features of the buildings invite or prevent crime?

Web project. Find Web sites that describe streetcars and trains from the past, how they were used, and what they did to daily life.

NOTE

1. This resulted from young people leaving home earlier to live in their own apartments, higher divorce rates, divorced persons living in their own places rather than going back to parents, separations, and elderly people living on their own. Sub-households also declined in relative numbers.

⋈ FIVE ⋈

MARKETING STOLEN GOODS

A thief can do this with the loot:

- Use it himself.
- Trade it for other contraband.
- Sell it directly to others.
- Sell it to a go-between, who sells it to the public.

If the thief[1] uses the stolen item himself, he has to like that particular CD, or the clothes must fit. If the thief trades it off, he has to find someone who wants it and has something to trade for it. If he sells it for cash, he has to wait around or find a buyer. The easiest solution is to find a go-between to take the loot off his hands and pay him quickly. That shifts over the slower process— selling—to someone else.

But the go-between does not pay well, so the thief may have to steal $100 worth of goods to get $5 or $10 cash. This formula explains the thief's problem:

The thief's "take" = the original value of the loot

   minus the markdown for damage while stealing it

   minus the markdown because it is used

   minus the markdown because it is stolen

   minus the markdown for the rush to get rid of it.

87

As the farmers say, it mostly goes to the middleman. Sutton's (1995, 1998) classic research finds three types of stolen goods handlers: those who fully know the goods are stolen (a "fence"), those who half know, and those who really do not know (see also Sutton, 2004; Sutton, Schneider, & Hetherington, 2001; Tremblay, Clermont, & Cusson, 1994).

The thief can try to get around stolen goods handlers by just stealing cash. But victims often watch their cash more closely than they watch their goods. Some thieves try to bypass the fence or other go-between by setting up a table at the flea market or going around selling the stuff. Yet the whole idea of crime is to get quick money for less work. Go-betweens keep crime moving, with important help from the public itself.

THE THIEF AND THE PUBLIC

When selling stolen goods, a thief depends on the public. People must be interested in the loot. Do not dismiss that interest as secondary. In fact, evidence is growing that the demand for these goods drives property crime, more than the other way around.

The Wheel of Theft

Thief and public are linked in a market, even if it is not quite the same as the market for toothpaste. Exhibit 5.1 depicts a very special sequence for crime, the Wheel of Theft.

1. The thief steals the loot from the public.

2. Then sells the loot back to the public.

3. Then gets money for it.

4. Then spends that money as a consumer.

The money flows clockwise in the inner circle, whereas the loot flows counterclockwise in the outer circle. Exhibit 5.1 neglects go-betweens for stolen goods, who could be slipped into the eight o'clock position. We could envision an extra wheel depicting the market for contraband purchased by the offender with his crime profits. Indeed, we can envision a series of wheels, like

Exhibit 5.1 The Wheel of Theft

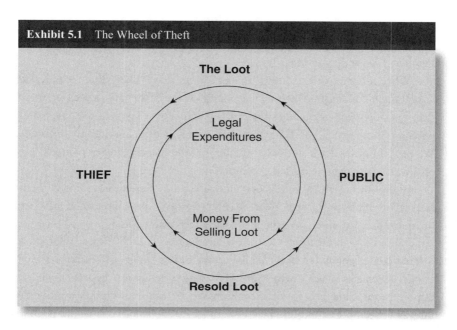

in an antique clock, with illegal goods and services being traded or spun in an intricate web of shady activity. That web gives an impression of organized crime, except that not all participants need to meet in a single place. The wheel of theft helps you understand the semivisible hand of criminal behavior and its significant consequences for society.

The Public Interest

The Wheel of Theft shows how much the offender depends on the public, and the public draws from the offender. Finney, Wilson, Levi, Sutton, and Forest (2005) provide empirical evidence for this symbiosis based on the British Crime Survey 2002–2003. Some 21% of the public admitted being offered stolen goods in the previous 5 years. Given that some people would probably not admit this, and that many stolen goods are not known to be so, the public likely purchase many more stolen goods than they admit or even realize. About 35% said they thought their neighbors' homes contained "a lot" or "quite a few" stolen goods, 52% thought "not very many," and only a minority (13%) thought their neighbors had no stolen goods at all.

Roughly 13% of males 16 to 25 years of age admitted having bought stolen goods. A much higher percentage (41%) said that someone had offered them such goods. Again, because people tend to avoid admitting unsavory facts, the real numbers are surely higher. Some surveys also indicate higher numbers than these.

When offered stolen goods, males are no more likely than females to accept them—at least by their own account. On the other hand, males are more than twice as likely as females to *receive* illicit offers. Thus, in the end, males purchase more stolen goods without showing signs of being morally different. We need to always ask not only whether people respond to temptations, but also whether they are exposed to temptations.

Eighty-three percent of respondents saying they were offered stolen goods also reported having turned down the offer. However, the more offers they get, the more they are likely to accept. Those offered stolen goods *once* resist the temptation 91% of the time. Of those offered *a few times,* 80% resist. Of those offered stolen goods *often,* 80% also resist. Stolen goods offers drive stolen goods purchases. That's why Sutton's "market reduction approach" is so important for crime prevention. Shrinking the stolen goods market would give thieves fewer outlets and less incentive to steal. Their markups would decline and they would have to work too hard at crime to make it pay. They would be turned down more often as they tried to unload the loot and would have to take additional risks of being turned in.

This British survey cleverly asked which of the goods purchased in the prior year were new versus used. They asked about purchases via small ads, the Internet, informal markets, pubs, homes, secondhand shops, and the like. Some 20% of laptops or other electronic equipment and 16% of bicycles were bought on the side. Whether the buyer knows it or not, many of these goods were probably stolen. Buying "on the side" is, however, a broader issue than stolen goods themselves.

Second-Class Goods

Interestingly, goods not stolen can help camouflage those that are. A modern society, with its tremendous production of goods, produces a vast array of second-class goods. These include

- Overstocked products
- Items no longer fashionable
- Slightly damaged goods

- Irregular items ("seconds")
- Legal imitations of popular items[2]
- Sellouts from stores that went out of business
- Used goods

A small store selling legitimate but second-class items can easily mix in some questionable merchandise bought on the side. Secondhand shops, small ads, and flea markets include legitimately used or old merchandise, with stolen stuff mixed in. It is no accident that a secondhand market in Europe was historically called a "thieves' market." Stolen goods can easily be dumped into this sea of legitimate second-class merchandise and sold within the secondary market. That's why the line between a fence and a legitimate merchant is often thin (see Cromwell et al., 1991; Johns & Hayes, 2003; Klockars, 1974; Steffensmeier, 1986; Sutton, 1995; Tremblay et al., 1994).

INVITING PEOPLE TO STEAL MORE

Stunning research by James LeBeau and Robert Langworthy shows us how society can produce more crime (Langworthy, 1989; Langworthy & LeBeau, 1992; LeBeau & Langworthy, 1992). Police departments in various cities can set up "sting operations" in order to catch thieves. The method is simple: open up a fake storefront to buy stolen goods, then arrest the offenders. It's very easy because the offenders come to you. Fence stings tend to bring the police excellent publicity and meet with widespread public approval (Newman, 2007).

LeBeau and Langworthy are trained geographers, so they know clever ways to study crime locations. In a real city with a police sting to fence stolen auto parts, they mapped exactly where cars were attacked. They found compelling evidence that the police had increased auto theft around the sting storefront by making it easier for offenders to get rid of the loot. The police had inadvertently harmed the very public that was cheering them on. They should have realized that theft thrives on the opportunity to unload the loot. Providing a convenient fence is probably one of the worst ideas that law enforcement has ever come up with. To their credit, some police departments figured this out themselves and decided to stop.

IT'S EASIER TO SELL
STOLEN GOODS TO THE POOR

Why do low-income areas have more crime, even when their residents are not especially criminally inclined? As Chapter 4 indicated, the answer lies in the tangible features of life for the disadvantaged. Nowhere is this more evident than in their interest in second-class goods.

More Frequent Temptations

Sutton's (1998) work found that those in low-income areas were about as likely as anybody else to *turn down* offers of stolen goods. But they were *offered* stolen goods more frequently. Respondents were divided into six areas by socioeconomic status; those living in the upper-class area were offered stolen goods only 7% of the time. Those in the lowest-class area were offered stolen goods 17% of the time. What they say about their neighbors may offer a better indicator. Only 9% of those living in the highest-status areas believe that "quite a few" or "a lot" of their neighbors have stolen goods in their homes. For the lowest socioeconomic group, 38% think that.

To interpret these data as proof that poor people are more "criminally inclined" completely misses the point. Poor people are clearly more inclined to favor used and secondary merchandise for economic reasons. Add this to the points made in Chapter 4, and you now have a much better understanding of why low-income areas can have higher crime rates without extra crime motivation.

Poverty Areas Offer More Outlets for Second-Class Goods

The consequence is that poverty areas have more outlets for second-class goods. And second-class goods provide more chances to unload stolen goods. As you drive through the low-income swath of a city, you may notice stores selling used appliances; pawnshops; sidewalk merchants; and signs stating "discounts," "seconds," "reductions," "liquidation," or the equivalent words in other languages. A city provides additional outlets for used merchandise, parts, or commodities including scrap rubber dealers, auto parts shops, auto body shops, retread tires, scrap metal dealers, car wrecking services, used car lots, junk dealers, and plumbing fixtures. There may also be stores advertising "We buy gold and diamonds." Many low-income areas also have street merchants offering used goods to crowds on the street, selling every kind of used product from an old fan to an old book or cheap chipped plates. As a result, items are stolen there that would be passed over in the suburbs. All of these could be outlets for stolen goods or provide means for camouflaging these goods among other items. It does not really matter who knows the goods are stolen; the point still holds.

The car lots in the central city sell only used models. Older cars need more replacement parts, which then become more valuable. That makes the cars targets of car-part theft while giving their owners an incentive to steal parts or to purchase those stolen by someone else.[3]

Even if some middle-class people would like the savings from buying second-class goods with their low markup, secondary stores cannot easily afford to move into a higher-rent district. Low-income areas solve the problem by offering low-rent storefronts. Exhibit 5.2 puts all this together, showing how a low-income area can pull in crime by providing both buyers of second-class goods and places to sell them.

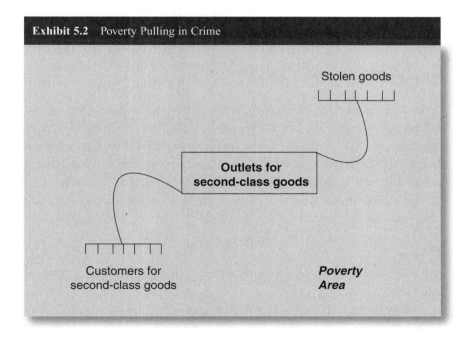

Exhibit 5.2 Poverty Pulling in Crime

As Chapter 8 will explain, one crime can lead to another. The *Poverty-Area Crime Sequence* helps explain how fencing opportunities can bring more than property crime to low-income urban areas:

1. The poverty area provides customers and outlets for second-class goods.

2. That encourages property offenders to be active in a poverty area.

3. Proceeds from these property offenses assist drug purchases and drug sales there.

4. More serious offenses follow and proliferate in the poverty area, derived from its stolen goods market.

Chapter 4 indicated many other features of poor neighborhoods that would make them crime prone; this chapter shows how markets for stolen goods can enhance those problems.[4] You can see that middle-class people have made a major contribution to the "moral order of suburbs" by refusing to buy second-class things from a shabby store or car lot with no warranty offered. If the United States, like Japan, exported all used cars after 3 years, we would save ourselves a lot of crime.

NEW AVENUES FOR STOLEN GOODS

As the routine activities of life change, so do the opportunities for buying and selling stolen goods. The Internet has changed the face of a lot of our every-day behavior, such as researching a half-forgotten fact or newspaper story, looking up a business in the Yellow Pages, and printing maps and directions to a friend's party. It has also affected stolen goods markets. Sites like eBay, Craigslist, and literally hundreds of others serve as the middleman for fencing stolen goods. The trend is so new that there is little criminological research about it, but some scholars have termed this phenomenon *e-fencing* (Hollinger, Langton, & Adams, 2007). Web sites are used to advertise and sell personal and other used or new items with no questions asked and almost no personal interaction.

Many of the Web sites sell products to the highest bidder, whereas others exchange goods, such as gold and jewelry, through the mail for cash. The anonymity of the Internet and the fact that thieves can often get 70% of the retail value of the product make this a very lucrative way to fence stolen goods (Sherman, 2009). The anonymity also attracts buyers who would not likely be offered stolen goods in the streets of low-income areas. Over the Internet, buyers are actively seeking "a good deal"—in some cases, not even knowing they are buying stolen property.

Thinking back to the crime triangle, in this situation, the Internet and these Web sites are the "places" where the process is initiated and the U.S. Post Office or other mail delivery service completes the transaction. Regulating these practices is very difficult because the Internet is so new and expansive, and it has no clear structure or ownership. However, in recent years, several legislative bills have been introduced at the federal level to address these issues. One of these bills is called the E-Fencing Enforcement Act of 2009. The bill seeks to require online auction operators to retain information about high-volume sellers and be able to provide that information if a crime is suspected (Sherman, 2009).

CONCLUSION

The market for stolen goods is very important, fitting within a larger market for second-class goods. Recently, it has begun to change with the increased use of the Internet. Once more, you can see that crime feeds on a larger system of activities. You can also recognize a *system* of temptations: to buy fancy legal goods; to buy second-class goods for a good price; to avoid asking tough questions when someone offers you a good deal; to steal the goods directly. A larger market links these temptations and their fulfillment.

How life tempts people, much more so than human character, helps you understand behavior. In the 1960s, the older generation disparaged younger generations for taking too many sexual liberties. The Illinois Reminiscence Survey (see Felson & Gottfredson, 1984) studied how the older generation itself behaved, and to what temptations it was exposed. Respondents were asked to think back to age 17 about a variety of activities, many linked to sexual behavior (see Chapter 6). An unpublished section of the survey found that the older generation was indeed somewhat less likely to report sexual experience by age 17. But those who claimed a lack of early sexual experience also reported that *nobody had ever asked*. Given a chance to engage in sex, the older generation said yes 90% of the time. The younger generation simply got more offers than their parents or grandparents. Opportunity makes the lover, just as it makes the thief.

This chapter has shown you that offers of second-class merchandise at low prices and availability of the Internet facilitate "regular folks'" participation in crime. The next chapter considers how crime opportunities contribute to the criminal behavior of youths.

MAIN POINTS

1. Thieves depend greatly on the general public for unloading the loot. Used goods and other second-class merchandise are important for camouflaging stolen goods.

2. The Wheel of Theft shows how much the offender depends on the public and the public draws from the offender. The thief steals the loot from the public, then sells the loot back to the public, then gets money for it, then spends that money as a consumer.

3. In the British Survey 2002–2003, one fifth of the public admitted being offered stolen goods in the previous 5 years. A majority thought their neighbors' homes contained at least some stolen goods. Men tend to buy stolen goods more than women, and men are offered stolen goods more often. These goods frequently include electronic equipment and laptops. Most people who say they were offered stolen goods reported having turned down the offer. However, the more offers they get, the more they are likely to accept.

4. Second-class goods often help camouflage stolen goods. Examples include overstocked products, items no longer fashionable, legal imitations of popular items, irregular items ("seconds"), sellouts from stores that went out of business, slightly damaged goods, and used goods.

5. Police stings that use fake storefronts to buy stolen goods can create crime that otherwise would not have occurred, by creating the opportunity for easily selling the stolen goods.

6. Poor people are more inclined to purchase second-class merchandise and to find it available on the street. They are more frequently exposed to stolen goods through flea markets and secondhand stores that are more likely to be in poor areas.

7. The Internet has created an entirely new and expansive system for selling stolen goods.

8. The process for marketing stolen goods is much more important than previously thought, could be the driving force for property crime, and has an indirect impact on other types of crime.

PROJECTS AND CHALLENGES

Interview project. Interview someone from the wholesale or retail goods sector. Without talking about crime, find out how and why goods are rejected by or shipped away from first-line stores and where they end up.

Media project. Look for advertisements in the newspaper or on TV that could conceivably involve stolen goods. What makes you suspicious of some but not others?

Map project. Pick a part of a city. Map out an area containing several possible outlets for stolen goods, even if the buyers and sellers might not know they are stolen. Show where the outlets are on the map and discuss.

Photo project. Photograph six store signs that seem to invite thieves to unload stolen goods. Discuss.

Web project. (On the Internet, find five Web sites that conceivably could be used to sell stolen goods. What about the way the Web sites work would make it easy to conduct this type of crime? Discuss.

NOTES

1. This chapter uses the word *thief* but really includes *any* offender unloading stolen goods. That might include burglars or even those who take jewelry or other goods in a robbery.

2. Of course, counterfeit items can also be sold as new, defrauding customers (see Chapter 7; Cell A2 of Exhibit 7.3 treats this as one form of white-collar crime).

3. Middle-class areas have some outlets for stolen goods, too. Research by Biron and Ladouceur (1991) at the University of Montreal found that shops that rent out VCRs to middle-class people also bought used ones from local teenagers, no questions asked. Stores in middle-class areas might also buy used jewelry.

4. Although a low-income area may provide residents access to stolen goods at a lower price (see Johnson et al., 1985, Chapter 12), victimization undermines these "benefits," giving residents a net loss in prosperity, not to mention their other risks.

SIX

CRIME, GROWTH, AND YOUTH ACTIVITIES

T he onset of puberty brings on many changes besides reproductive capacity and interest in sex. It also increases stamina, visual acuity, activity levels, and physical size. For males, puberty leads to rapid acceleration of growth, especially muscle (see Felson & Haynie, 2002; Meschke & Silbereisen, 1997; Montemayor, Adams, & Gullotta, 1990). Thus, adolescents are ideally suited to physical activity, for better or for worse. Of course, they are also suited for sexual intercourse and everything related to it.

For thousands of years, Jewish law stated that a girl becomes a woman as soon as she grows two pubic hairs. In the United States, however, we view teenagers, even college students, as preadult. Our society has created a fundamental mismatch between adult abilities (including sexual capacities) and adult designation. That leaves modern youths without a satisfactory position in society and undermines society's ability to keep them out of trouble.

YOUTH AND CONVENTIONAL ROLES IN THE PAST

The young spent most of human history better suited to conventional roles than they are in America today. Full of energy, strong, and sexually capable, they were physically prepared for working hard and raising a family.

Almost every activity once required the heavy use of muscle: farming, hunting, chopping wood, shucking corn, scrubbing clothes, weaving baskets, building wagons, shoeing horses, churning butter, raising cattle, or forging iron (see Wigginton, 1972, 1979, 1999). Youths had an important economic function, with productive tasks to burn off energy. They needed little or no schooling; in 19th-century America, most instruction ended by age 12 so the young could get to work. Teenagers were ready to marry young: females by 16 and males by 18. If babies did not come quickly, something was wrong.

Exhibit 6.1 compares the main stages and timing of a girl's reproductive life in this earlier era compared to now. A girl was menstruating at about age 14 and she would marry at 16. If she started intercourse right after puberty, she probably would not get pregnant until age 15, allowing a 9-month cushion to set up the wedding (no limousines needed). The boy was working anyway. Thus, there was little conflict between the end of schooling, sexual maturity, sexual formalization, and work responsibility. As long as the boy was not a slacker and took the vows, the older generation did not care much that the seed was germinating. They needed the young men and women around for farming, factory work, homemaking, and construction. Young adults were expected to get to it.[1]

Exhibit 6.1 Main Stages of a Girl's Reproductive Life, Earlier Centuries Versus Today

Age	Earlier Centuries	Today
12		Onset of puberty
13		Body ready for pregnancy
14	Onset of puberty	
15	Body ready for pregnancy	
16	Ready to marry and give birth	
17		
18		
19		
20		
21		
22		Ready to marry

THE CHANGING POSITION OF YOUTH

A major transformation occurred in the 19th century in the United States and numerous industrial countries affecting the roles of youth and hence their crime and delinquency.

Biosocial Change in Adolescence

The protein-rich diet of modern society led to a remarkable physical change (see Montemayor et al., 1990). In the 20th century, puberty most commonly began around age 12. These averages do not tell the story for every individual, but they provide an idea of what to expect for the population as a whole. This change in the onset of puberty has profound implications for both the productive and the reproductive roles of teenagers, because work roles shifted in exactly the opposite direction.

Changes in the American Occupational Structure

The era of butcher, baker, and candlestick maker is long gone, giving way to tremendous specialization. The butcher of today includes such occupations as animal stunner, shackler, sticker, head trimmer, carcass splitter, offal separator, shrouder, hide trimmer, boner, grader, meat smoker, and hide handler. Even computer programmer now can be disaggregated into hundreds of specialties. Specialized jobs for a complex workplace require more years of schooling and experience. Mechanized work no longer makes use of young muscle, endurance, or sensory acuity.

The Combined Effect of Biosocial and Occupational Change

Lacking a suitable economic function, the young must live for the future, studying, working part-time, and responding to sexual urges on an ad hoc basis. With physical adulthood at 12 or 13 and full adult roles at 22 or later, the young have to fill the many years between (see Exhibit 6.1, right column). Although young people today tend to commence sexual activity by age 15, that is actually a longer wait after puberty than may have occurred in the 19th century. For society to delay an urge as powerful as sex for 3 years or longer is truly a remarkable accomplishment. Economics is more powerful than tradition:

If a girl today gets pregnant at 15 or 16, few 21st-century parents want the young man to hang around their daughter for life.

The 1950s—The Calm Before the Storm

The transition toward greater crime and delinquency was somewhat delayed in the 1950s. Even though younger puberty and later adult roles were well on their way, the post–World War II baby boomers were not yet teenagers. Most mothers with young children were not working full-time. When school was over, teenagers were subject to some supervision because mothers were still present in the neighborhood. That left a short window of opportunity for getting into trouble. Suburbs did not have enough shops to steal from. Electronic goods were still too bulky to steal easily. The multicar family was unusual.

The Life of Teenagers Changed Very Quickly

From the 1960s onward, the life of teenagers was transformed, and crime opportunities with it. In particular, the shift of women into the labor force left residential areas largely unsupervised in the afternoon. Even if one teenager's mother was around, mothers of friends would be away for the day. That opened up a vast opportunity for teenagers to do whatever they wanted.

The Illinois Reminiscence Survey documented dramatic change in adolescent activity patterns over several generations (see Felson, 1998; Felson & Gottfredson, 1984). Telephone interviewers contacted a sample of the Illinois population aged 18 and older, asking respondents to think back to age 17 and describe their routine activities in some detail. They asked very explicit questions about who was home in the afternoon, including the following:

- When you were 17, between 3:00 p.m. and 6:00 p.m. on weekdays, were you at home always, most of the time, some of the time, or never?
- When you were at home between 3:00 p.m. and 6:00 p.m. on weekdays, was an adult there with you?
- On weekdays, did you have any regular household chores or duties to perform in the afternoon?
- Did all the members of your household usually eat the evening meal together on weekdays?

For both males and females, the answers to these questions shifted greatly from generation to generation. Those reaching age 17 in the 1960s and 1970s were much more likely than previous generations to escape parental controls in the afternoons. They were home less often, had parents around them less, and had fewer afternoon chores to keep them occupied. Perhaps the most interesting finding was that having the family dinner together, nearly universal in the older group, declined noticeably for the younger group. Today, you can probably ask your fellow students and find an even more extreme change, with very few reporting having routine weekday family meals together.

Suppose two teenage boys want to get together to avoid parents—come what may. The right circumstances are more likely to occur if neither one has afternoon chores or has to be home for family chores. The major historical decline in risk of parental interference with teenage plans has an important impact on youths, making it easier for them to converge and act independently of parents (Felson, 1998). As society turns them over to an adolescent world, all they have to worry about is one another.

ADOLESCENT CIRCULATION AND CRIME INVOLVEMENT

Extensive transportation systems increase risks that young people will be involved in crime and delinquency, whether as offenders or as victims.

Public Transit and Adolescent Involvement in Crime

Teenagers in cities use public transit more than middle-aged people. That is especially true for low-income youths living in areas dominated by automobiles. Also, young people old enough to get into trouble, but too young to drive, might well use public transit. By exposing youths to strangers, public transit can increase their victimization risk, as well as their offending. Public transit also helps youths escape adult supervision. Some cities even provide a youth pass with unlimited travel. Some young people jump subway turnstiles or sneak into the back doors of buses, getting into trouble afterward. Some teenagers in Vancouver have used the Skytrain, a monorail, to go to the mall parking structure, where they break into cars, and then go into the mall for more offending (Brantingham, Brantingham, & Wong, 1991).

Transit stations and their vicinities in Europe and the United States are often crime settings for youths, including young prostitutes (Jansen, 1995; van Gemert & Verbraeck, 1994; West, 1993). Teenage runaways were very common inside New York City's Port Authority Bus Terminal, often selling sex or carrying out other "hustles" (Felson et al., 1996; see also Chapter 9, under the heading "A Huge Bus Terminal Turns the Corner"). Taking advantage of commuter crowds, youths are also involved in panhandling, selling or purchasing drugs, stealing luggage or other items, attacking the homeless, or whatever else they might think up. Many are themselves victims of predatory crime (see Baron, 1997; Baron, Forde, & Kay, 2007). Indeed, public transit can simultaneously move guardians away from residential areas while making them targets elsewhere, increasing crime risk throughout the metropolis.

Autos for Youths

An automobile intrinsically provides substantial freedom of movement to its driver. A youth with a car could break laws more efficiently and over a wider area. The Illinois Reminiscence Survey once more showed, from the middle of the 20th century, a dramatic change in the mobility of youth. Most noteworthy were increased driving after dark and riding around with other teenagers.

In the 1950s, when families typically owned a single car, their teenage drivers were very low on the pecking order to use it. Parents might drop them off and pick them up, but that meant they knew more about where their children were and what they were doing, and they had better ways to check the information. In the United States today, well over half the households have two or more motor vehicles. This is very important for young drivers. Two-car families more than double adolescent access. The parents might go out on a weekend evening in one car, leaving the other car under full teenage control.

In the car-dependent metropolis, bus service gets weaker, especially on evenings and weekends when social life is greatest. Declining densities of cities make walking less efficient. Heavy and fast automotive traffic is no friend to pedestrians or bicycles. As cars become virtually the only feasible option for getting around, youths who do not drive are increasingly dependent on parental chauffeuring. This gives parents a very strong incentive to provide their adolescents with cars. Increased adolescent mobility reduces their supervision in a quantum leap.

Adolescent Convergent Settings

An emerging topic in crime is the study of co-offending; that means committing crimes together—something especially common among teenagers (Warr, 2002). For that to happen, offenders must converge socially or otherwise to find partners in crime. The process of co-offending (Felson, 2003) is a serious factor in generating more crime or less. One of the central questions is how school itself may contribute to criminal acts among youths by bringing them together.

SCHOOLS AND CRIME

The school is the heart of the adolescent weekday circulatory system. It pumps adolescents into society around 3:00 p.m., then pulls them back in the next morning.

Timing Is Everything

By assembling lots of youths, then dumping them simultaneously, the school sets the stage for quite a number of problems. Exhibit 6.2 shows dramatic changes in violent incidents hour by hour in the course of a school day versus a nonschool day (see Snyder & Sickmund, 2006). Between 3:00 p.m. and 3:59 p.m. on school days, about three times as much aggravated assault victimization is reported as between 1:00 p.m. and 1:59 p.m., whereas on nonschool days, only 1.4 times as much aggravated assault victimization is reported between 3:00 p.m. and 3:59 p.m. as between 1:00 p.m. and 1:59 p.m. The data also show a noteworthy increase from 2:00 p.m. to 2:59 p.m. on school days, perhaps reflecting early releases, and perhaps indicating that some youths left early on their own authority. From 4:00 p.m. to 6:00 p.m., there is still plenty of trouble on school days, but it declines as the evening proceeds and goes down to its lowest points after midnight. On non-school days, aggravated assault peaks at 8:00 p.m., and then significantly declines as the evening progresses.

School Location Is Very Important

School location also influences crime rates. City blocks containing a secondary school have higher crime rates (Roncek & Lobosco, 1983; Roncek &

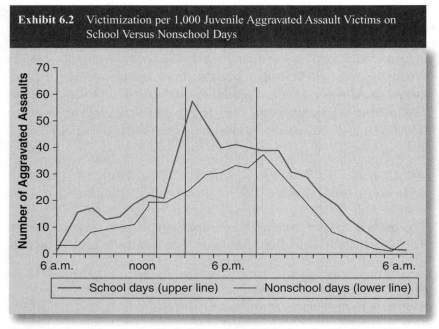

Exhibit 6.2 Victimization per 1,000 Juvenile Aggravated Assault Victims on School Versus Nonschool Days

SOURCE: Adapted from Snyder, H., & Sickmund, M. (2006). *Juvenile Offenders and Victims: 2006 National Report* (p. 34). Pittsburgh, PA: National Center for Juvenile Justice.

NOTE: The x-axis depicts a 24-hour timeline, beginning with 6:00 a.m. The first line to the left denotes 1:00 p.m. The middle line refers to 3:00 p.m., and the right line is at 8:00 p.m.

Maier, 1991). In personal communications, Patricia and Paul Brantingham have explained the importance of the routes that young people take to and from school. If those routes pass by good crime opportunities, you can guess what happens. Placing a high school and shopping mall side by side proves a disaster for both. The merchants feel that the students make a mess, steal from the stores, spend too little money, and drive out paying customers. The educators know it's hard to learn algebra at the mall.

Secondary schools test our crime analysis to the fullest. On one hand, they assemble youths at prime offending and victim ages, then dump them en masse. On the other hand, they provide a degree of supervision and hence crime prevention. As much as we complain about schools and note the problems within, much more trouble occurs outside their span of control and after school is over. Jackson Toby (1995) argues against requiring teenagers to stay in school against their will, mainly to improve education for those who want to be there. We argue for keeping them in school, where they find relatively fewer

- Valuable crime targets
- Chances for arguments to escalate
- Partying opportunities
- Conflicts with other citizens

When schooling is required, any youth away from school on a school day will draw suspicion. Of course, localities with staggered school hours assist local truancy and delinquency; any youth asked why he or she is not in school can simply answer "I'm off today."

You can see that the school role in crime prevention is highly routine. Don't be taken in by the terrible incidents, such as school shootings. Most school-related crime consists of small thefts and truancy. The risk of violence per hour spent inside during school hours is quite small, and the aggression that does occur is usually quite minor and self-contained. As you can see from Exhibit 6.2, most of the risk that young people will commit crime occurs near the hour that school lets out—especially when young people are on the way home and no longer under adult supervision (Dinkes, Cataldi, & Lin-Kelly, 2007; Felson, 1998; Gregory, 2000; Lab & Clark, 1997; New Jersey Legislature, 2005; Snyder & Sickmund, 2006; State Education Department, 1994). This evidence flies in the face of most of the publicity about violence *inside* the school during school hours.

THE CENTRAL ROLE OF SCHOOL SIZE

High crime rates in or near schools are not inevitable. It has long been known that smaller secondary schools tend to have significantly lower crime rates than larger schools (Garbarino, 1978; Gottfredson, 2001; Gottfredson & Gottfredson, 1985; Hope, 1982; State Education Department, 1994). Secondary schools exceeding 1,000 students generate about 60% to 70% higher crime rates than schools with 500 students or less.

Why Large Secondary Schools Generate Extra Crime

With larger numbers, students are less likely to know one another and administrators less able to tell who belongs there. Burrows, Shapland, Wiles, and Leitner (1993) interviewed principals (head teachers) in British schools, most with fewer than 800 students. Some 65% said they would be able to tell whether a person entering school property was a stranger or not. An American principal in a school of 2,000 or more students and 75 or more teachers could not easily match this (see Dedel, 2005).

School size undermines adult control of the premises in another interesting way. Suppose that 1% of all students are the most difficult to control. In a school of 600, that makes only six problem students. The teachers can "divide and conquer" to minimize the harm they do. In a school of 3,000, 1% makes 30 problem students; together, they can bully the rest.[2] The important issue is school size, not classroom size.

Small schools also engage a higher percentage of students in extracurricular activities, even if large schools have more offerings. Barker and Gump (1964) demonstrated that small schools obtain significantly higher levels of student participation. The reason is that their limited teams, clubs, and classes are short of participants and need to recruit those of average talent to join.

Size of Building, Grounds, and Greenery

Parents are easily impressed by spacious grounds. But trees and greenery do not make adolescents good. British researcher Tim Hope (1982) demonstrated that they instead contribute to higher school crime rates. Trees provide good cover for sneaking away or getting high. Even though the modern American school has many more teachers, it has far fewer teachers per acre to watch over space.

Parents seem to like campus designs and to dislike the old red brick schools of the past. But those schools offered more control and security. Students compacted into a single building of three stories, situated on a small cement lot, were under greater adult control. Bullies had a harder time taking over. The relatively few halls and stairwells were easy to watch. The principal's office, located strategically by the first-floor door, allowed monitoring of who entered or exited.

To sum up, Americans in the 20th century enhanced school crime inside and out by giving up small schools. That served to reduce participation in school activities, make school social functions too large and risky, and undermine adult supervision, resulting in few educational gains for most students.[3] This is not to say that smaller schools are perfect in every way. You can bet that the choir will be off-key, the basketball team will make fewer three-pointers, and the baseball team will drop more balls; but at least all participants will get their chances and do so rather safely.

PARENTAL TRIALS AND ERRORS

Parents and other adults keep trying to adapt to the changing position of youths. In general, they have sought to organize recreation and work for young people in order to keep them out of trouble. But these efforts have often failed or even backfired.

Physical and Outdoor Recreation for Youths

After-school and weekend recreation should, in theory, keep young people out of trouble. With this in mind, during the period 1900 to 1920, many cities constructed parks, playgrounds, and youth centers. Labor was cheaper then, permitting hiring of playground supervisors to prevent trouble. In recent decades, costs of labor rose and municipal budgets declined. As playground supervisors faded into history, parks were increasingly taken over by the nastiest youths. Parents might drive their children somewhere for organized recreation once a week, but ongoing daily recreation lacks such controls.

The pecking order for control of urban space follows a pattern something like this:

1. Strong adults

2. Nasty teenage boys

3. Other teenage boys

4. Teenage girls

5. Younger children and weak adults

When the strong adults are removed from the scene, the second group on the list takes over the parks and playgrounds. The challenge for a modern urban society is to figure out how to empower the first group to look after recreational space.

With school over at 3:00 p.m. and working parents returning at 6:00 p.m. to 7:00 p.m., an afternoon vacuum of at least 3 hours remains. Even if school activities keep a young person occupied for an extra hour, two thirds of the afternoon vacuum remains unfilled. Parents who lack the money to pay for extra activities will not find it easy to fill this vacuum. With nobody home to cook dinner, a family meal together is impractical.

Those parents who can afford it or make an extra effort might be able to keep their children occupied for longer hours in organized activities. This can be very important and effective, but it is easier to promise than to carry out consistently. Recreation programs, no matter how wholesome and educational, can backfire by assembling likely offenders who might not have gotten together otherwise.

Recreation programs should be careful to provide sufficient supervision to cover the period from 3:00 p.m. to 7:00 p.m. These programs should inform parents fully what supervision is provided and what is not. They should tell parents directly when football practice is cancelled; otherwise, teenagers will use that time to fool their parents into thinking they are being supervised. They should make sure that activities do not dump groups of youths on the street together after the adults are gone. Youths finishing football practice should never be released in a pack. Planning only part of the afternoon vacuum leaves the rest of it available for delinquency. On the other hand, after-school recreation can provide reduction in youth arrests (Gottfredson, Gerstenblith, Soulé, Womer, & Lu, 2004).

Some adults have extremely naïve ideas about sporadic recreation. There was a program in Los Angeles that took adolescents to the movies once a month, claiming to be reducing crime and delinquency. What about the rest of the month? A parent who wishes to program a teenager's activities for the month does not have that easy a task. The period from 3:00 p.m. to 7:00 p.m., 20 days a month, totals 80 hours. Add to this the difficulty of influencing later hours and you can see that parental control in modern American life is dubious indeed. A divergent metropolis makes this supervision quite difficult by providing many settings where young people, using cars, can assemble largely free from parental influence. In comparison, European societies provide transportation systems that assemble large numbers of youths in predictable central city areas, so their control problems—although serious—are not as dispersed as in the United States.

Youth Recreation in an Electronic Age

The age of speedy Internet communications provides new options for youths to break laws, often operating out of their homes. They can produce their own pornography. They can view pornography by others. They can sell themselves as prostitutes. They can make sexual liaisons with those their own age or ages well beyond their own. They can send and/or receive threats via the Internet and buy or sell contraband goods. They can, at a young age, learn how to hack the computers owned by others or distribute computer harm in various ways. They can participate in cyber chat rooms to discuss all of this. A few criminologists have begun to study these behaviors among juveniles (Finkelhor, Mitchell, & Wolak, 2000; Patchin & Hinduja, 2006), and law enforcement is increasingly interested in adults who take advantage of youth involvements.

The use of cell phones and text messaging further enhances the ability of youths to evade adult controls. Youths can easily meet, finding the niches when they are not being supervised and quickly determining where friends are located. Thus, it becomes harder and harder for parents to keep them under wraps. Researchers are only beginning to figure out what all this means for crime.

The Impact of Part-Time Jobs

Both intuition and basic control theory tell us to keep young people busy. Work seems like a good way to do that. Most of us were taught that work builds character, so jobs for youth really ought to reduce their criminal behavior.

Researchers have disappointed us by finding out what we do not want to hear: High school students who have jobs get involved in more crime, not less (Apel, Bushway, & Brame, 2007; Brame, Bushway, Paternoster, & Apel, 2004; Fagan & Freeman, 1999; Gottfredson, 1985; Greenberger & Steinberg, 1986; Runyan, Bowling, & Schulman, 2005; Staff & Uggen, 2003; Wright & Cullen, 2000). Why does the logic of control theory not hold for teenage jobs in today's society? Consider the past and present characteristics of teenage work. In the past, prosperity levels were low and work requirements time consuming and physically demanding. A 16-year-old would work to help himself and his family survive. He or she would already be married, or marriage would be

imminent. When work was over, there was sleep. For youths working as farmers and in family businesses, parents were very near. Young factory workers
found themselves among older workers, who would tend to keep them in line.

Modern prosperity and late marriage combine to change the significance
of teenage jobs. The typical 16-year-old worker does not spend the money to
support self or others. Earnings go for recreation and extras, including illegal
drugs, underage alcohol, or car expenses that help in evading parental controls.
The work itself places limited physical demands and leaves many youths full
of energy after finishing their jobs. The low-wage and low-skill jobs available
to youths often are performed near other youths and with few adult workers
around. Many jobs (e.g., in retailing) bring the worker closer to money and
goods suitable for theft.

You can see why getting jobs for youths to thwart their crime and delinquency is so likely to backfire. Again, finding a suitable role for young people
in modern society is quite a difficult challenge.

Modern Packaging of Alcohol Makes It Less Controllable

Many modern products make it easy for young people to escape parental
controls (Felson, 1998). In recent decades, the packaging and marketing of
alcohol has changed dramatically in the United States. Whiskey is increasingly
sold in large bottles (e.g., 2 liters). As parents overstock their liquor cabinets,
their children can readily siphon off alcohol without drawing parental attention. Who will notice an inch missing from a very large bottle? Who would
notice a bit of watering down? The dispersion of liquor consumption away
from bars and toward package sales also changes the dynamics of alcohol control: A single identification card—real or fake—can get a dozen underage
youths extremely drunk. Adolescent use of cars makes it easy to consume alcohol away from parental supervision, often in the cars themselves.

CONCLUSION

Many features of the divergent metropolis help involve teenagers in crime
and delinquency. In general, modern society leads to less control. It is a mistake to interpret this as a "cultural change" or "moral deterioration." Rather,
it represents a shift in the tangible features of everyday life. Bodies, products,

technology, and transportation have also changed, undermining adult control over teenagers.

MAIN POINTS

1. A major transformation occurred in the 19th century in the United States and numerous industrial countries affecting the roles of youth, and hence their crime and delinquency.

2. Biosocial change in adolescence came from a protein-rich environment that shortened the onset of puberty and had profound implications for both the productive and the reproductive roles of teenagers, because work roles shifted in exactly the opposite direction.

3. Changes in the American occupational structure to highly specialized jobs requiring more years of schooling and experience means that young, strong adults have fewer opportunities for work.

4. Changes from the 1960s onward, in particular when more women went to work, meant that the lives of teenagers were transformed with the lack of supervision.

5. Extensive transportation systems increase risks that young people will be involved in crime and delinquency, whether as offenders or as victims.

6. Modern life puts young people in a bad position, taking away their historical roles in work and family life. However, it puts them in a good position for escaping parental supervision.

7. Public transit exposes youths to strangers, which can increase their victimization risk, and it also helps youths escape adult supervision.

8. Automobiles greatly enhance the ability of youths to escape parental controls. That causes crime to disperse over a wider area, while providing good targets for illegal attacks.

9. Schools help to control crime and victimization of youths, but they also facilitate problems by bringing many youths together and then releasing them simultaneously at particular times and in particular areas.

10. Managing the time of teenagers becomes a major problem for society. Many ill-considered and ineffective control methods are still offered and believed effective by people who think little about the timing and location of adolescent activities.

11. Undersupervised outdoor and indoor recreation, part-time jobs, and packaging of alcohol are significant contributors to youth offending.

PROJECTS AND CHALLENGES

Interview project. Interview three high school students about how they use text messaging and cell phones to escape parental controls.

Media project. Look at media treatment of young people as offenders 20 years ago and today. Has it changed?

Map projects. (a) Map a secondary school and major nearby housing areas. Emphasize the main paths those teenagers on foot take home in the afternoon. (b) Map an entertainment area or district where young people often go. Map bars, attractions, hangouts, and so on. Predict trouble spots.

Photo project. Photograph a path often taken by teenagers. Take a photo every 10 feet; array these photos into a sequence with discussion. Note any litter, vandalism, or signs of burglary. Note where problems have not occurred.

Web project. (a) Find Web sites that deal with school bullies and bullying. What do they say about dealing with the problem? Is their case persuasive? (b) Check the Web site of the Memphis-Shelby Crime Commission (2000) for its after-school recommendations. How do these fit concerns expressed in the current chapter? (c) If you are a user of a social networking Web site, think about the dangers of a youth sharing too much information about him- or herself and accepting strangers as friends. How can parents help prevent and monitor their children's activity on these sites?

NOTES

1. But do not be nostalgic for the past, with its low standard of living, difficult work, epidemics, and infant mortality.

2. See Baldry and Farrington (2007); Espelage and Swearer (2004); Greif and Furlong (2006); Junger-Tas and van Kesteren (1999); Kepenekci and Cinkir (2006); Olweus (1993); Phillips (1991); Pitts and Smith (1995); Smith et al. (1999); Smith and Sharp (1994); and Tatum (1993).

3. Felson (1998) explains why small schools have more and better interracial contacts—because the critical number of each group is small and interactions are more natural. Of course, larger schools give a better superficial appearance of racial integration because their numbers look good.

SEVEN

WHITE-COLLAR CRIME

———•◦•———

I f you go back to Chapter 1, you will understand why "white-collar crime" is such a confusing topic.

Recall the dramatic fallacy—the tendency to focus on the most dramatic crimes. In the area of white-collar crime, it's easy to pay attention to those who deal with a billion dollars, while forgetting how often people take small and medium amounts from their employers and customers. Most occupational illegalities are "relatively mundane, unskilled, easily accomplished, and modest in economic return" (Wright & Cullen, 2000, p. 863).

Note the cops-and-courts fallacy, namely, overstating the importance of the justice system. In the "white-collar" crime field, cases that receive little or no prosecutorial attention are often neglected by analysis.

The ingenuity fallacy means assuming that offenders are very clever, when in fact, they are usually very ordinary. Most white-collar offenders are far from geniuses, stealing while nobody is looking (Benson & Simpson, 2009; Weisburd, Waring, & Chayet, 2001).

Finally, the agenda fallacy warns us to beware of crime analysis linked to moral, religious, social, and political agendas. Many of those who emphasize white-collar crime are very politicized. But in this book, we try to find a calmer approach to crime analysis.

Indeed, most white-collar crime need not involve white-collar workers at all. Any employee at any level might be stealing from the company or its customers, or abusing its clients. Is it really so fancy to pilfer accounts, make off with equipment, shortchange customers, write yourself a check, spill secrets, promote a secretary for sex, or endanger others by snoozing on the job? Yet these mundane offenses are important to society precisely because they are widespread and very costly.

White-collar offenses are difficult, if not impossible, to count with any accuracy. Crime rates based on Uniform Crime Report data from the FBI do not even consider these crimes. Yet the many categories of white-collar offending discussed in this chapter cost at least tens of billions of dollars. Moreover, they undermine public trust and harm society beyond the money lost.

White-collar crime also lays down the gauntlet for crime theorists. Michael Gottfredson and Travis Hirschi (1990) have challenged us to state a single theory that could handle all crime under one roof. Yet efforts to bring white-collar crime into the larger field have only begun. One major problem is that the term itself is far too vague. That ambiguity makes it hard to grapple with its nature and causes. Some of the leading writers on "white-collar" crime call their own field an "intellectual nightmare" (Geis & Meier, 1977, p. 25), in a "state of disarray" (Wheeler, 1983, p. 1655), with "more confusion than ever" (Friedrichs, 2006, p. 5). The purpose of this chapter is to reduce that confusion by considering the practical routines that generate these offenses. Recently, Weisburd and his colleagues (2001) suggested that white-collar crime is largely understandable in terms of routine activities and crime opportunities. Our theoretical task in this chapter is to show how.

Past scholars of white-collar crime failed a simple test: They could not find a couple of simple sentences to give clear definition to their category. Such a definition must put *like* crimes together and set them apart from *unlike* crimes. At the same time, the definition must not put this category on a different planet from other crime categories. But a definition of crime cannot be based on a vague image of social groups.

AN IMAGE OF A CRIMINAL
GROUP IS NOT A REAL DEFINITION

It does not make sense to simply define a crime category according to a social category of offenders. Unfortunately, powerful imagery guides how many people think about crime. As Exhibit 7.1 indicates for the United

Exhibit 7.1	Images of Crime Versus Reality

	Popular Images		
Race	Economic Image	Crime Image	Crime Reality
Black	Poor	Violent offenses	Humble offenses predominate
White	Middle class	Extreme "white-collar" offenses	Humble offenses predominate

States, popular images have black people being poor and committing violent offenses. Meanwhile, white people are supposed to be members of the middle class and to carry out extreme white-collar offenses. Those having strong political or social sympathies can use these images for propaganda. If you are unsympathetic or patronizing toward black people, you can think of them as violent. If you dislike business or majority power, you can focus on white-collar crimes or use the second image to offset the first.

There's just one problem: *Both* crime images are inconsistent with reality. The same sources documented in Chapter 1—official statistics, victim surveys, and offender self-reports—show that offenders in all racial and socioeconomic groups commit mainly humble and ordinary crimes, such as shoplifting, burglary, driving while drunk, running red lights, stealing little things, or smoking small amounts of marijuana. As explained earlier in this chapter, plain pilfering and finagling in common occupations greatly overshadow "crime in the suites." The last column of Exhibit 7.1 shows us the humble reality for offenders rich and poor, white and black.

Here you see a classic example of false images created by adding percentages the wrong way—something you learn to avoid in the first week of statistics class. A high percentage of people in trouble for fancy crimes are indeed white; but a very low percentage of white offenders are in trouble for fancy crimes. Black offenders might be overrepresented in armed robbery, but that offense is a tiny share of criminal acts, or of their misbehavior. Most black *and* most white offenders are mundane.

Defining white-collar crime as elite crime makes absolutely no sense in today's world. A large part of the difficulty is that white-collar workers today are a good majority of the labor force, so we cannot simply equate "white-collar crime" with "crime in the suites" or offenses committed by those in elite ranks.[1] The finagling we associate in our minds with white-collar crime is not really that different from tricks by blue-collar workers, such as those auto mechanics who charge for services not rendered, industrial workers who pilfer at the plant, or restaurant workers who scam the customers or owners (Benson & Simpson, 2009; Blumberg, 1989; Crofts, 2003; Greenberg, 2002; Grey & Anderson-Ryan, 1994; Kidwell & Martin, 2005; Locker & Godfrey, 2006; Payne & Gainey, 2004; Rickman & Witt, 2003).

If most offenses by white-collar employees have standard motives and methods (see below), why do we find it necessary to distinguish white-collar crime from general fraud or pilfering? Friedrichs (2006) suggests that white-collar offenders are different because they are "trusted criminals," yet the same could be said of many rapists or non-white-collar tricksters. Weisburd, Wheeler, Waring, and Bode (1991) emphasize that these are "crimes of the middle classes," but middle-class people also offend beyond the boundaries of what we consider white-collar crime—burgling, driving drunk, abusing drugs, and beating wives. Smigel and Ross (1970) provided a catchy and yet focused term, "crimes against bureaucracy." That, however, omits many offenses normally classified as white collar, such as cheating clients or dumping toxic wastes. So we are back at square one.

A MATTER-OF-FACT DEFINITION

Never define hot dogs by the mustard they go with; instead, visit the meat-packing plant and learn how hot dogs are produced. Likewise, never define a type of crime by the *people* who might do more of it, least of all the color of their collars or skin. Instead, check the specifics of what the offenders do. Then figure out what these specific illegal actions have in common.

In his thorough and interesting book, *Trusted Criminals: "White-Collar" Crime in Contemporary Society,* David O. Friedrichs (2006) gives us a good place to start. Friedrichs discusses scores of white-collar crime types occurring in various professional, corporate, and government settings. If you study his list, a single general feature of white-collar offenders leaps out: Each one had specialized access to crime targets via a particular work role, profession, or organizational position.[2]

That generalization covers such diverse offenses as fixing prices, short-changing customers, cheating pension funds, faking insurance claims (see Gill, 2000b), fiddling expense accounts, dumping toxic wastes, abusing prisoners, mistreating patients, manipulating stocks, and enticing bribes. Indeed, a specific work role can put someone in just the right position to do the wrong thing.

Here, we rename the category "white-collar crime" as *crimes of specialized access*. Occupations, professions, and organizations provide offenders with practical routes to their targets. The central point is that *legitimate* features of the work role provide a chance to do misdeeds. Knowing the passwords helps in breaking into accounts. A stockroom job brings a chance to pilfer. Responsibility for control of checks gives a chance to misuse them. The inspector can solicit bribes. Work roles and positions in organizations explain all these crimes and more. The final definition, simply stated, is as follows:

> Crime of specialized access: a criminal act committed by abusing one's job or profession to gain specific access to a crime target.

With this definition, a crime of specialized access always has a target, such as money, and at least one victim who loses that money or something else of value or importance. The public at large could be victimized, but even here the crime of specialized access is a predatory act.

So defined, crimes of specialized access are an identifiable category with a clear rule for determining what fits and what doesn't. This definition incorporates a vast variety of crimes commonly called white-collar crimes, including offenses by those higher and lower on the totem pole. Many corporate crimes are also contained within this category, as long as individual culprits are identifiable.[3]

DO CRIMES OF SPECIALIZED
ACCESS HAVE SPECIAL MOTIVES?

The scores of white-collar offenses presented by Friedrichs (2006) suggest many motives, most commonly material gain in some form. Many offenders might have been driven by such motives as power, sex, friendship, approval, excitement, revenge, avoidance of retaliation, or evasion of chores. But this list could just as well fit shoplifters, burglars, or vandals. Violent offenders, too, pursue money, power, sex, revenge, or self-protection (Tedeschi & Felson, 1994).

Thus, in some ways, white-collar offenders fit the same social mold as burglars, robbers, thieves, and assailants. Their motives do not set them apart from other criminals. Of course, many of the cases tried in criminal courts define a corporation as the defendant. Yet to analyze these cases in practical terms, we must focus on the individuals who did the dirty work. Even when a corporate interest drives a white-collar crime (Paternoster & Simpson, 1993), individuals must be motivated to comply with that interest.

Here we find one of the most irritating features of traditional treatment of white-collar crime: the notion that it is something fancy rather than something simple and crass. To make progress in studying these crimes, we need to follow the same principles we use for other offenses: Learn the specifics, find out the modus operandi of the offenders, search for the practical features of the criminal acts, and find out what the offenders want. Among the obstacles to understanding crimes of specialized access has been confusion about who commits them. Interestingly, these offenders range all the way from teenagers whose jobs give them access to crime opportunities (Wright & Cullen, 2000) to the infamous spy and FBI turncoat, Robert Philip Hanssen. Both working teens and tricky spies use their legitimate roles to carry out illicit acts. In that sense, despite their criminal motives, they use special methods. They rely on the specialized access provided by their work roles. Such access does not necessarily require high rank.

Organizational Rank

Most of us feel intuitively that those higher on the totem pole would have relatively more criminal access than those lower. However, high positions do not always bring good crime opportunities. Meanwhile, some lower- or middle-tier jobs offer excellent criminal access. These could include

- A janitor who uses his keys to steal equipment
- A hotel desk clerk whose inside knowledge helps him rape a guest
- A police officer who solicits bribes for looking the other way
- A purchasing agent with easy access to kickbacks

A good research agenda would try to find out how each work role sets up crime opportunities. Such research may, in time, show us that all occupational ranks provide some willing offenders and some crime chances. Within each rank, crime chances will vary with duties, shift assignment, and supervision. A good position for carrying out one type of crime might be little help for doing another.

Crime chances might shift greatly as a person moves about in a career or an organization. Imagine an offending employee moving up the job ladder. She starts her first year as an assistant cashier, stealing $1,000 from the register. The second year she is promoted to the counting room, slipping away with $3,000. The third year she works as dock supervisor, taking away goods with a retail value of $10,000, but a fence value of less than $2,000. In the fourth year, she is chief accountant in the main office, writing checks that transfer $10,000 to her own uses. She was *willing* to steal at all levels, but found somewhat greater illegal opportunities as she went up the hierarchy.

In general, we expect that, for crimes of specialized access,

- Any given individual will be able to do more harm with higher rank.
- Being more numerous, low-ranking positions together add up to more total harm.
- Employees harm employers more than the other way around.

Empirical research is needed to verify, disprove, or modify these hypotheses and to teach us how each occupation provides access to crime opportunities. Which positions handle money? How much and in what form? Exactly what jobs have access to computer accounts, files, keys, passwords, or confidential information? Which positions control or disperse valuable resources to others? Which employees or managers make decisions impinging on the health or welfare of others? Which jobs can deliver favors that might draw bribes? We have an ample research agenda for crimes of specialized access.

HOW SPECIALIZED ACCESS PERMITS CRIME

Fortunately, crimes of specialized access lend themselves particularly well to the routine-activity theory applied throughout this book. Routine official duties provide the opportunity to misbehave unofficially. The following section explains how this fits within a larger system of activity.

All Crimes Need Access

Suppose you return from a trip out of town and find that your home has been burgled. You think the offender knew you were away, but are puzzled how he might have found that out. Your puzzlement points right to some central issues in studying crime.

Every predatory offender must solve a basic problem: to gain access to crime targets. Three general methods are available to help the offender solve that problem (Exhibit 7.2). First, the offender might take advantage of *overlapping activity spaces,* perhaps learning about your impending absence by seeing you pack your car or leave in a taxi for the airport. Second, the offender could rely on *personal ties,* such as hearing about your trip through mutual friends or neighbors. Third, the offender could use *specialized access* to information about you, such as a postal clerk handling your request that your mail be held in your absence or the taxi driver who takes you to the airport.

Exhibit 7.2 shows us that white-collar offenders have found a *particular solution to the offender's general problem.* That solution, like the other two, is based on routine activities, as crime carves its niche into everyday life.

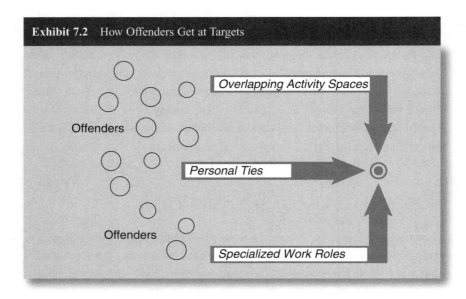

Exhibit 7.2   How Offenders Get at Targets

Offenders

Offenders

Overlapping Activity Spaces

Personal Ties

Specialized Work Roles

Some Offenses Can Be
Committed Only Via Specialized Access

Most of the offenders committing Part I crimes (see Chapter 1) probably get to their victims via overlapping activity spaces or personal ties, not specialized access. But the exceptions (such as the postal or hotel clerks mentioned earlier) impel us to find a way to distinguish a coherent category of specialized-access crime.

Some offenses can be carried out *only* through specialized access (or, at least, usually so). These offenses help us delineate a fairly distinct class of crimes, also fitting fairly well the traditional list of white-collar offenses.[4] Yet the great variety of such offenses demands some further organization.

Specialized access allows harm to victims in four ways. First, offenders can make *illicit transfers* of money, goods, or other resources to the detriment of others. Second, they can abuse their access to *misinform* others. Third, they can apply their specialized roles to *manipulate* others. Fourth, they can *endanger* the health and safety of others.

Victim Types and the Specialized-Access Crime Grid

Type of harm and type of victim are presented in the Specialized-Access Crime Grid depicted in Exhibit 7.3. The grid contains 72 examples, some more common than others. The first row of the grid outlines how *employees* might be victimized. The boss could steal their payroll deductions, conceal discrimination, manipulate employees into illegal acts, or endanger them through illegal overtime work.

The second row of the grid describes harm to *clients,* including customers and patients, sometimes the direct result of company policy. Included are plundering trust funds, violating privacy commitments, tricking people into signing a document, or failing to warn someone of risks in the products they are buying.

The third row considers crime against the *general public.* It includes cases in which a company diverts public property to its own use, falsifies its financial reports to mislead investors, fools its government regulators, twists research to avoid bad media coverage, or corrupts outside auditors.

Exhibit 7.3 The Specialized Access Crime Grid

The Victims	*How to Abuse Your Position and Harm Others*			
	A. Transfer	*B. Misinform*	*C. Manipulate*	*D. Endanger*
1. Employees	Cheat overtime Ransack pension funds Defraud payroll Underpay credit card tips	Distort worker rights Scandalize insubordinates Hoodwink unions Fake nondiscrimination	Fix union election Rig promotions Pressure for sex Enlist workers as offenders	Imperil workplace safety Misuse child labor Hire labor racketeer Coerce long hours
2. Customers, Clients, or Patients	Pass counterfeit goods Resell stolen goods Plunder trust funds Cheat at register	Overbill services Misrepresent products Use phony diploma Violate privacy	Fix customer prices Take gift for referral Sell by trick Finagle fine print	Sell risky product Abuse sexually Botch operation Fail to warn
3. Public	Receive stolen goods Cheat on dividends Divert public property	Falsify stock reports Fool regulators Twist research	Fix market prices Bribe public officials Corrupt auditors	Spoil environment Imprison unjustly Conceal perilous errors
4. Own Organization	Fudge timecards Abscond with equipment Embezzle services	Pad expenses Spread product hoaxes Double-cross partner	Take kickbacks Promote for sex Sell on the side	Sabotage project Neglect machines Protect negligent work
5. Other Organizations	Steal secrets Swindle software Defraud accounts Evade taxes	Falsify deliveries Cook up insurance claim Trick suppliers Claim bogus bankruptcy	Pay kickbacks Restrain free trade Delude inspectors Shift blame	Train unsafely Sabotage competitor Understate danger Shift danger to others

NOTE: Real life is hard to classify, no matter how good your categories. Several of the criminal acts listed above cannot be contained entirely by a single box.

Organizations might spoil the environment by dumping toxic waste or evading pollution controls. Public agencies can also endanger the public by violating civil liberties, imprisoning people unjustly, or mismanaging corrections facilities. In addition, any organization that covers up its previous errors might risk the health or safety of its clients. An example of this is the Firestone-Ford scandal over dangerous tires on certain vehicle models.

As shown in row four, *one's own organization* can fall victim to various crimes of specialized access. Employees who fudge their timecards or embezzle services (such as long-distance calls) are harming their own organization. This includes employee theft and many forms of crime at work detailed in research by the Scarman Center, at the University of Leicester in England (e.g., Bamfield, 1998; Beck & Willis, 1995; Gill, 1994; see also Dabney, Hollinger, & Dugan, 2004; Hollinger, 1993, 1997; Langton & Hollinger, 2005). Sometimes, one business partner will double-cross the other to maximize his own profit. A very common and important offense is taking kickbacks from outside vendors. For example, a purchasing agent for your university might accept a fancy gift after steering a contract to an inferior vendor. The boss who promotes a subordinate for sex (instead of quality work) is harming his own organization's success. Some manipulative employees sell their company's goods and services on the side. Those who fail to inspect machinery or protect against negligent work can endanger people and employer alike.

Finally, some organizations victimize *other organizations,* including competitors, suppliers, and governments. Examples are stealing secrets, evading taxes, selling flawed products, or tricking other companies out of goods or money. Large companies might force smaller competitors out of business, in violation of antitrust laws, or shift blame for their errors to another company (as Firestone and Ford did to one another).

Future Research

The Specialized-Access Crime Grid helps us assemble a vast array of offenses carried out via work roles. The purpose of the grid is to organize thinking and guide future research. By displaying transgressions by victims and workers alike, this grid helps the student resist "blame analysis"—a tendency to label groups you like as the victims and groups you dislike as the criminals (R. Felson, 2001; R. Felson & S. Felson, 1993). As you proceed, ask yourself these four basic questions:

1. Exactly what did the offender do?

2. With whom did the offender do it?

3. Who were the victims?

4. How did specialized access make the crime possible?

You now have a clear method for comparing and contrasting these offenses with other criminal acts.

CONCLUSION

Crimes of specialized access, whether in suites or shops, can be defined and studied as a part of ongoing life. These offenses fit within larger crime analysis. These, like other offenses, involve selfishness in its many forms. As with other predatory offenders, access to crime targets is preeminent. To analyze these crimes, the modus operandi is critical for our understanding. As always, illegal acts must carve their niche within a larger system of legal activities. Indeed, we can understand how these crimes change and vary only by studying the structure of legal activities in everyday life.

MAIN POINTS

1. "White-collar crime" is poorly named, because any work, any professional or occupational role, can get involved. These offenses are usually not dramatic or ingenious.

2. These are better called "crimes of specialized access." These crimes depend on individuals being in a position to do a particular harm.

3. Offenders of these crimes range from teenagers whose jobs give them access to crime opportunities to corporate executives and high-level government officials.

4. It is a myth that race or economic standing predicts whether individuals will commit violent crimes or crimes of specialized access.

5. White-collar offenders have found a particular solution to the offender's general problem of gaining access to crime targets. Of the three general methods that are available to offenders—overlapping activity spaces, personal ties, and specialized access—white-collar offenders use the last.

6. Some offenses can be carried out only through specialized access (or, at least, usually so). These offenses help us delineate a fairly distinct class of crimes, also fitting fairly well the traditional list of "white-collar" offenses.

7. These offenses might harm employees, one's clients, the public, one's own organization, or other organizations.

8. These offenses fall into four categories: Offenders can make illicit transfers of money, goods, or other resources to the detriment of others; they can abuse their access to misinform others; they can apply their specialized roles to manipulate others; or they can endanger the health and safety of others.

PROJECTS AND CHALLENGES

Interview projects. (a) Interview any person in any occupation about minor crime at the workplace. Ask only about general awareness to avoid incriminating anyone. (b) Pick any offense in Exhibit 7.3 and interview somebody about how this would be done. Does this offense belong in its current location in the exhibit? Where else might it be placed?

Media project. Compare news media coverage of "white-collar crime" cases to the empirical work in Weisburd et al. (2001).

Map project. Map a business or plant where you have worked. Figure out the different types of crimes of specialized access that can occur within it, or, using its location, note how the physical layout of the place contributes to crime opportunities.

Photo project. Get permission to photograph some local business or professional offices, avoiding the names or signs. Imagine what crimes of specialized access could conceivably occur at each.

Web project. Identify a Web site and think about how it could be used to make illicit transfers of money, goods, or other resources; misinform people; manipulate people; and/or endanger the health and safety of people.

NOTES

1. When Edwin Sutherland coined this term (Sutherland, 1939), white-collar occupations consisted of a small share of the labor force. The elite image made more sense then. But changes in society have made Sutherland's definition largely obsolete (see Weisburd et al., 2001; Wright & Cullen, 2000).

2. The word "specialized" is used here rather than "special" because the former has a work and professional connotation.

3. This definition also includes a subset of organized crimes—those that occur through legitimate activities. For example, organized crime took control of the New York docks via the men who worked there and the owners of various companies (Bell, 1960). However, drug sellers in the park and their suppliers are acting entirely outside of legitimate roles and thus cannot be included in this definition.

4. A subset of offenses uses specialized access to reach targets. A subset of these uses specialized access exclusively. A small subset of these involves conspiracy in any sense, and a very small subset of these involves organized crime in the full sense.

EIGHT

ONE CRIME LEADS TO ANOTHER

———————•◆•———————

C rime has multipliers. Sometimes, violence breeds violence. Sometimes, burglary leads to fencing stolen goods. Sometimes, a minor amount of alcohol gets someone into big trouble. One crime can feed into another. Our task, if we decide we can handle it, is to figure out just how.

The bad news in this chapter is that one crime often multiplies harm over time. The good news is that reducing crime in one way may help reduce it in another. In the past few decades, we have seen both vicious and auspicious spirals, reflecting the interdependence among crimes. We will examine how one crime is linked to another on five levels: the offender, the market, the victim, the offense, and the setting. After briefly discussing the offender's crime interdependencies and then the interplay of illegal markets, most of the chapter considers the latter three means for one crime leading to another.

THE OFFENDER'S SLIPPERY SLOPE

Criminologists have long known that committing one offense correlates with committing another. Crime has its pushes and pulls on offenders. An offender's illegal act can encourage subsequent offending in many ways. It can provide money to buy drugs, feed dependencies, reinforce future illegal efforts, set up time in risky settings, expose the offender himself to victimization, require self-defense, and draw retaliations. The more we study offenders, the more we learn that crime is a slippery slope toward more crime.

Yet the active offender sometimes desists (Cusson, 1990, 1993). As he moves along the life cycle, perhaps engaging in prolonged substance abuse, he may experience illness and injury, premature aging, loss of jobs and families, bodily deterioration, even physical incapacitation or death. A heavy criminal lifestyle eventually undermines crime itself. At that point, the offender may spend more time in the hospital than enjoying his loot or his intoxicants. However, this may not happen until late in the criminal game.

THE INTERPLAY OF ILLEGAL MARKETS

One crime leads to another, too, because illegal markets interact. For example, prostitutes often use or sell illegal drugs. May, Hough, and Edmunds (2000) have explored the interplay of sex markets and drug markets, showing that sex workers purchase drugs in significant amounts and support the drug market as retail workers. Other illegal markets are probably interlinked, too. Some illegal gambling operations might assist sales of illegal drugs. Handlers of stolen legitimate goods might also sell counterfeit labels. After-hours bars and others violating liquor licenses might sell drugs. Illegal prostitution can depend on markets in smuggled or even kidnapped women. Illegal drug sales in large quantities create a need to launder money. Selling illegal goods spawns buying illegal goods. Although markets for illegal goods and services sometimes help people avoid direct participation in crime, markets can also involve people in more illegal activity.

EVEN VICTIMIZATION SPAWNS MORE CRIME

In the past, we did not normally think of victimization as a cause of crime, but we know better now.

Repeat Victimization: A Growing Field of Study

Ken Pease (1992, 1997) has shown us how important repeat victimization really is. A burglar may break into your house in September, wait until you have replaced the goods in October, then break in during October to remove them again. Many shoplifters hit the same stores over and over. We can see that each offense helps generate the next. Ken Pease (1992) and his colleague Graham Farrell (1995) call this "once bitten, twice bitten." Bullying is a form of repeat victimization because the victim is often the same. The study of repeat victimization is an important growth field in the study of crime and its prevention (for a review of repeat victimization literature and analysis techniques, see Weisel, 2005).

The Victim Becomes the Thief

One of the more subtle and interesting facts about crime is that a victim becomes—soon thereafter—a thief. Jan van Dijk (1994) is a leading criminologist in the Netherlands, where bicycles are commonly owned, ridden, and stolen. For many Dutch people, a bike is a necessity for everyday transportation; almost everyone knows how to ride a bike. It is not uncommon for someone whose bike was just stolen to steal someone else's. This sets in motion a *Van Dijk chain reaction,* as depicted in Exhibit 8.1. First, A steals B's bike. Then B steals C's bike. Next, C steals D's bike. Finally, D steals E's bike, but E is left out in the cold. This happens, too, in entertainment districts. One person comes out of the bar to find his bike gone, so he then steals someone else's, and so on. Although it's a simple example, the important point is that each crime can generate many more.

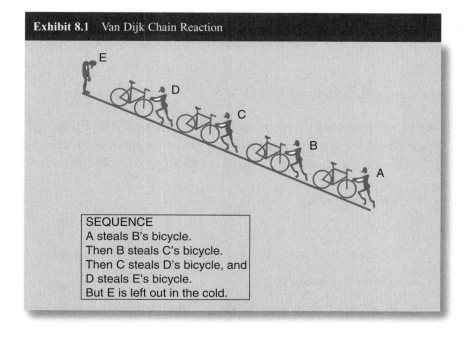

Exhibit 8.1   Van Dijk Chain Reaction

SEQUENCE
A steals B's bicycle.
Then B steals C's bicycle.
Then C steals D's bicycle, and
D steals E's bicycle.
But E is left out in the cold.

Van Dijk chain reactions occur in the United States with certain auto parts or car stereos. Other chain reactions can be set in motion when a worker picks up a tool belonging to another worker. If you look around in your own environment, you may see that van Dijk chain reactions are common at universities, where bicycles, required textbooks, and student computers disappear regularly.

Crime Repeats by Block and Time Slot

Sherman and Weisburd (1995) translated parts of the routine activity approach into the term "hot spot." A hot spot is a location or area containing much more than its share of crime or disorder events. One example is a study by David Weisburd and his colleagues (Weisburd, Bushway, Lum, & Yang, 2004) in Seattle, Washington. They examined crime along the city's 29,000+ street segments, and over a 14-year period, they found that about 50% of the total crime occurred within about 5% of the street segments.

Crime events also cluster in time (Groff, 2007; Grubesic & Mack, 2008; Ratcliffe, 2004; Robinson, 2008). You can expect barroom problems to repeat on weekends. Problems on the way home from school repeat daily when high schools close. Daytime burglary recurs in the middle of weekdays. The progression toward Christmas annually leads to extra property crime and crimes linked to making merry with the assistance of alcohol. The advent of spring brings with it a growth in many types of crime.

QUICK LINKS AMONG OFFENSES

The circumstances created by one offense can set a quick stage for another. Remembering the stages of crime discussed in Chapter 2, this is explained by *criminal event chains,* with one criminal event leading to another, as follows:

1. The prelude situation leads to

2. incident A, followed by

3. the aftermath of incident A, which creates the prelude situation for

4. incident B, followed by

5. the aftermath of incident B, which creates the prelude situation for another incident.

For example, a bunch of boys are hanging around. They decide to break into a nearby building. After doing so, they are confronted by someone who tells them to stop. They assault that person, and so on. As we learn the common sequences, we will be able to further corroborate the general point: One crime facilitates another.

Camouflage

Certain offenses provide camouflage for other offenses. For example, loitering, unlicensed street sales, blocking sidewalks, and panhandling are all illegal acts of a minor nature that also assist those who might hang around. These include drug sellers, prostitutes, and pickpockets. Or, marijuana sales might provide camouflage for selling harder drugs. These offenders can offer camouflage for those who would rob customers for illegal goods and services at gunpoint.

Crime Facilitators and Between-Crime Linkages

In Chapter 2, you learned that some *props* contained within a setting can make it suitable for crime. For example, a couple of angry guys might be more likely to commit aggravated assault if a weapon is handy. Ronald Clarke (1997b) has defined a broad category, "crime facilitators," taking into account anything that helps crime along. That could include tools, weapons, drug paraphernalia, or disinhibitors—such as drinking too much—whether at the potential crime scene or elsewhere. A crime facilitator can be present before, during, or even after a crime (consider an escape car or a toilet for flushing down illicit drugs).

Legislation and law enforcement often focus on controlling facilitators. Of course, society cannot ban flush toilets simply because drug dealers use them to dispose of the evidence. But society does try to limit access to dangerous and addictive painkillers, guns, explosives, drug paraphernalia, and store tag removal tools. Society also tries to keep spray paint out of the hands of juveniles, who often paint graffiti. The success in controlling facilitators varies greatly.

Many illegal acts are difficult to commit without also violating a law controlling its facilitators. This gives police an extra basis for arresting people. For example, opening a beer can inside the car crosses the line even before a drop is swallowed.

Vulnerable Targets

Finding a target for the first crime, the offender sees additional illegal opportunities. He breaks into a home, then finds a woman there to rape. This malicious serendipity is a common component of sexual assault (LeBeau, 1987; Warr, 1988). A burglar finds a business to break into, then engages in vandalism while he's at it. Even murderers have been known to pick the wallet off the corpse. A prostitute can slip the wallet from the unaware customer's pants dangling on the chair, or steer him into a robbery by her collaborator. What if a prostitute's customer refuses to pay, or if the two disagree on what he owes? They cannot turn to small claims court to settle their differences. As one judge said, "I do not run a collection agency for prostitutes."

Vulnerable Offenders

Offenders are vulnerable to their victims, too. Wronged victims might defend or counterattack. They could refuse to cooperate with offenders, forcing flight or escalation. Once co-offenders are conveniently assembled, they might as well commit another offense. They could find a new victim, but they could very well fight among themselves over the loot or what to do with it. One of them could rob the other. They could spend the money on more alcohol or drugs and hence commit additional offenses in the course of the evening.

Small traffic violations can lead to dangerous fights or predatory attacks, such as when one driver cuts off another, who then retaliates (on road rage, see Asbridge, Smart, & Mann, 2006; Britt & Garrity, 2006; Galovski & Blanchard, 2004; Garase, 2006; Harding et al., 1998). Indermaur (1995) has a wonderful title to his book, *Violent Property Crime*, reminding us not only that some property crime is implicitly violent, but also that much of it leads to unplanned violence.

Follow-Up and Cover-Up

An illegal act, no matter how small, immediately compromises the offender's position. Virtually all offenders need to cover up the first crime, often requiring them to commit a second one. A burglar might assault anyone who he thinks will incriminate him, even another offender. The simplest theft gone awry can lead to a struggle and even a serious assault. Even something so small as a traffic violation can lead the guilty party to speed away to avoid detection, then to be chased by police, then to be charged with resisting arrest.

The need to hide crime also exposes offenders themselves to extra risk of victimization. A respectable customer for a prostitute or drug seller is not likely to seek police help after being robbed. Nor is the female prostitute or male hustler likely to seek police assistance after someone seizes the money from illegal gains. Again, one crime leads to another.

CRIME LINKS IN LOCAL SETTINGS

We have seen how an offender digs himself deeper into crime, a victimization leads to more crime later, and an offense itself brings extra crime along quickly. Now we consider how a location sees crime grow.

Broken Windows and the Spiral of Decay

James Q. Wilson and George Kelling (1982) first developed the concept of "broken windows," which tells us that local deterioration and the proliferation of minor crime undermine a neighborhood.[1] This produces such a mess that major crime follows. This theory was credited with the turnaround in New York City's crime rates. Transit Police Chief William Bratton used these ideas to improve security on the New York City transit system (Chaiken, Lawless, & Stevenson, 1974). He later became police chief for all of New York City, where he put the same concept into practice (Bratton & Knobler, 1998).

Similar thinking and empirical evidence are developed by Wesley G. Skogan (1990) in his book *Disorder and Decline: Crime and the Spiral of Decay in American Neighborhoods*. Skogan has warned against allowing decay to proceed and has made it clear that minor and major offenses are intricately related on the basis of locality. According to Skogan,

> Once a community slips into the cycle of decline, feedback processes rapidly take control. The problems that emerge can include more serious forms of disorder, as well as escalating crime—consequences that further undermine the community's capacity to deal with its problems. (pp. 12–13)

This conclusion directly contradicts the finding that delinquency areas remained unchanged in crime, even as their population compositions shifted (Shaw & McKay, 1942). Skogan's Chicago findings were consistent with what

Leo Schuerman and Solomon Kobrin (1986) found as they tracked Los Angeles localities over a 20-year period. In the worst areas, crime shifted from being an effect of something else to becoming itself a cause of more crime. "Broken windows" and "spiral of decay" are very important insights, but they still need to be specified. Indeed, recent research is inconsistent in supporting these insights, but that is because the theories are not yet precisely stated. By removing some of the popular misconceptions about the theory and sharpening it up, we can also reduce the flip-flopping in empirical findings.

Zero-Tolerance Misses the Point

Broken windows theory is not about zero-tolerance by police (Kelling, 1999; Kelling & Coles, 1996; also see Burke, 1998; King & Brearley, 1996). All policing and crime prevention planning require discretion. The police cannot arrest everyone who goes 46 m.p.h. in a 45-m.p.h. zone. Herman Goldstein (1990), the father of modern police research, explains that police can design specific lists of their enforcement priorities, such as what noise levels are tolerated, priorities in arresting panhandling, and just what will not be tolerated in public places (see also Goldstein, 2003, 2005). That avoids the risk that vague "disorderly conduct" charges will either be abused by police or tossed out in court. By admitting that police must use discretion, and making explicit enforcement rules consistent with law, a community can reduce disorder systematically, legally, and fairly.

Beauty Is Not the Problem

It is wrong to assume that anything unattractive generates crime. Tim Hope's (1982) important British research found that well-maintained schools had less crime, but well-treed schools had more. Studies of crime control on the New York City subway found that preventing turnstile jumping minimized other crime, whereas reducing ugly graffiti had no such extra benefit (Chaiken et al., 1974; Sloan-Howitt & Kelling, 1997).

A general principle emerges: Beauty does not make people good. Indeed, lots of people enjoy getting dreadfully drunk in parks, listening to the songbirds. Residents of prim suburbs can still commit burglary or seek forbidden orgasms. Neat malls still have shoplifting. Attractive products are stolen all the more.

Nor does shabbiness necessarily make people bad. Urban villages with peeling paint are still secure (see Chapter 4). Some types of deterioration and debris *might* lead to crime, but only if they communicate clear loss of control over a thing or a place. For example, abandoned homes and factories are easy to take over for criminal purposes. The question is not whether something is pretty or ugly but whether the environment offers the offenders temptations or produces controls. Sometimes, temptations or controls have aesthetic correlates. Judge local crime vulnerability in very practical terms. Small problems can grow and annoy, but we have to think about how.

Nibblers and Grabbers: The Selfish Abuse of Public Space

When very few people do very little harm, a local area can handle it well. Park Slope, Brooklyn, allowed a single spot on Sixth Avenue for a village beggar. He always begged politely. Park Slope was a rather secure neighborhood, and his begging was beside the point.

But minor offenses can have major significance. In the *Yale Law Journal*, Robert Ellickson (1996) wrote a fascinating article about controlling panhandling and abuses of public space (see Ellickson & Been, 2005). He sums up the problem: "The harms stemming from a chronic street nuisance, trivial to any one pedestrian at any instant, can mount to severe aggravation. As hours pass into days and weeks, the total annoyance accumulates."

We would refer to these offenders as *nibblers*, because they nibble away at the public resource and make it less usable or bearable for everyone else.

Does a lot of minor crime really add up? The public widely agrees that a single aggravated assault with injury is far more harmful than one property incident. But what if there are a million of the latter? Borrowing the pieces from other research, M. Felson (1998) estimated that the total harm to society from theft and burglary is two to three times greater than that from aggravated assault. If all other minor offenses were included, the case would be stronger still. Could five million instances of aggressive panhandling harm the community as much as a million thefts?

Some urban offenders seize for themselves still larger slices of what belongs to the public. We call these *grabbers*. Imagine riding the subway and seeing a single person sleeping across four seats, making others stand. One person sleeping in the subway for hours could inconvenience hundreds of people.[2] Aggressive panhandlers are especially efficient at interfering with people. Those illegal street vendors, who place their wares just outside the subway leaving little room for the public, do much more harm than those who set up on a wider sidewalk.

It Depends on Where and When

Minor violations of law can be very important or trivial, depending on where and when they happen. To understand this, we have to put aside legal definitions of crime and consider where and when behaviors occur. The same basic behavior can have a far worse consequence in one setting than in another. Noisy dominance of the park is far more intrusive and destructive than a noisy party within a private area. Private or indoor drug sales do far less harm than public or "open-air" drug markets. Similarly, public prostitution is repulsive to families and intrudes on legitimate activities. A thoughtful prostitute will ply her trade by discrete advertisement and telephone.

Private drunkenness may annoy a few people, but public drunkenness widens the span of damage, drives people away, exposes those drunk to attack, and even ruins a neighborhood. In a small neighborhood pub, one can undermine one's liver among friends willing to offer a lift home. But large bars packed with strangers often generate more serious difficulties (Homel, 1997; Scott & Dedel, 2006). Even underage drinking is worse in a very large bar or nightclub dominated by youths and offering access to illegal drugs. Even worse is public drinking by crowds of young males—far more likely to generate fights and to disturb and threaten the larger community. Worst of all is being drunk within sight of small children, who tend to scare easily.

Incremental Loss of Public Space to Selfish Use

Outdoor offenders might try to see what they can get away with: first drinking in public occasionally, then hanging around longer, next taking over the corner, on to the park for a few hours in the evening, then taking that over entirely. The next thing you know, they run the whole area and do anything they want. Public parks then become inaccessible to most of the public; in effect, privatized by the toughest and nastiest guys. As offenders get out of hand, they overrun other youths, parents, place managers, and police. That's why "order maintenance" is an important part of policing.

It makes sense for police and other public officials to pay great attention to control of outdoor urban space, as long as they note exactly which of the small offenses truly do larger harm. This attention to daily detail may prove far more important than relegating police to the reactive role, where they consider mainly calls for service, crime after it becomes serious, and arrests late in the deterioration process. On the other hand, with the order maintenance model, police try to act early, before things get out of hand. They focus on minor crimes that are strategic and can lead to something worse.

Problem-Oriented Policing

Orlando W. Wilson was one of the most famous police chiefs in all of history. After a terrible scandal in the Chicago Police Department,[3] he was called out of retirement to clean up the mess. His top assistant was Herman Goldstein, now a professor at the Law School of the University of Wisconsin. Goldstein's excellent insight and practical knowledge led to a landmark book, *Problem-Oriented Policing*. Goldstein (1990, 1997) reminds us that the goal of police should not be to arrest people or gain convictions. Their purpose is to reduce crime. If one person fights another in a bar, the police officer should consider not just those two people but whether that bar creates ongoing problems.

Applying other research to this question, we know to ask, "Is the bar serving people after they are already drunk? Is the bartender inexperienced? Are the bouncers initiating violence? Are they letting underage youths enter illegally?" (see Scott & Dedel, 2006). With problem-oriented policing, the goal is to find the underlying problem and then seek a solution that will prevent several crimes from happening. For example, the police can talk to a particular bartender or bar owner about managing customers more effectively. Even better, they can work with the liquor board to apply better management techniques for several bars in town.

Some police officers are really good at involving local people in crime reduction. These officers save themselves and other officers a lot of trouble. Instead of arresting more people, sitting in the station to fill out paperwork, and going to court for hours on end, they just talk to people at the right time. Many bartenders and patrons will calm things down if asked. The tough police function is really just a backup for smarter efforts. Any police department relying on a nasty approach has already lost the consent of the public and will have a miserable time trying to use their own bodies to block crime.

In recent years, advocates of problem-oriented policing have created a virtual Center for Problem-Oriented Policing (POP Center) at www.popcenter.org. The goal of the center is to create, store, and distribute relevant research and practice on problem-oriented policing and, more generally, crime prevention strategies and research. The POP Center produces small booklets that summarize knowledge about crime and disorder problems, responses to these problems, and tools to employ to help understand problems locally. The guides draw on research findings and crime prevention practices in the United States, the United Kingdom, Canada, Australia, New Zealand, the Netherlands, and Scandinavia. Even though laws, customs, and practices vary from country to country, communities everywhere experience common problems. These guides bring this information together in a succinct way appropriate for police and undergraduates, and they are referenced throughout this book.

Importance of Civil, Administrative, and Traffic Law

Normally, crime makes us think of criminal law. But civil and administrative laws are often more important than any other aspect. Bars and taverns are regulated by liquor and licensing boards. Slum buildings with repeated illegal activities within are subject to civil abatement. Owners of empty buildings can be forced to board them up or convert them to proper use. Individuals can be given a ticket for leaving keys in the car. Parking enforcement can thwart certain types of drug sales or prostitution. Think about how much easier it would be to enforce civil and administrative laws against 30 owners than try to enforce criminal codes against 10,000 citizens. Recently, Mike Scott and Herman Goldstein (2005) wrote a guide for police, *Shifting and Sharing Responsibility for Public Safety Problems,* that outlines strategies for using these laws and engaging others besides police to address crime and disorder problems.

A breakdown in civil, administrative, or traffic law enforcement can lead to a proliferation of criminal acts. Unbridled pubs can have violence, even murders. Vacant properties can become drug dens, especially if allowed to sit there. A failure of governments to enforce their noncriminal laws might be more significant than their criminal process.

Routine Precautions

Ronald Clarke and Marcus Felson have defined the concept of "routine precautions" taken by citizens to prevent their own victimization (see

Felson & Clarke, 1995). A theft could serve as a "wake-up call," stimulating people to take routine precautions, perhaps reducing victimization risk for even more serious offenses. Sometimes, different types of offenses are in conflict with one another. Some shady businesses want the area safe for their own offending. Street prostitutes may chase away drug dealers or public drinkers who scare off their own customers (see Cohen, 1980; Skogan, 1990). Problem-oriented policing can help educate and reinforce routine precautions, making citizens more effective in their crime prevention efforts.

THE SYSTEM DYNAMICS OF CRIME

That illegal activities are interdependent is not very controversial; however, criminologists disagree on how that interdependence occurs.

The "Displacement Illusion"

For more than a century, many criminologists believed that the crime rate is a "basic social fact" that cannot be shaved away. They believed that profound changes in society and community were essential for crime rates to shift. We now know better. Two fundamental shifts in knowledge force us to discard that old idea.

As noted earlier in this chapter, local crime rates vary greatly from block to block and address to address. Thus, a socially homogeneous neighborhood can have dramatic crime variations *within* it, suggesting that very tangible and very local differences generate much of the variations in crime risk.

Crime prevented on one block or in a micro area is not simply displaced to other local areas. Crime prevented here will very often disappear, or at least most of it will not simply be pushed to a nearby place.

When Germany required that owners install steering wheel locks on all cars, old and new, car theft reduction was dramatic and was not displaced to other vehicles. Webb (1997) concluded that steering wheel locks had indeed reduced auto theft without displacement. Rene Hesseling's (1993) review of a large number of crime prevention efforts found that significant displacement seldom occurred, and even if it did occur, there was not a one-to-one level of displacement. That is, not every crime that was prevented in one setting was replaced by another crime in another setting; but if 10 crimes were prevented, and if displacement did, in fact, occur, fewer than 10 occurred in another setting.

Recent research by David Weisburd and colleagues (Weisburd, Wyckoff, & Ready, 2006) reinforces our conclusion that local crackdowns and crime prevention efforts produce a net reduction in crime. The crime is not simply pushed elsewhere. These results have been confirmed by other research (Bowers & Johnson, 2003; Braga, 2005; Brantingham & Brantingham, 2003; Cohen, Gorr, & Olligschlaeger, 2007; Lawton, Taylor, & Luongo, 2005; Ratcliffe, 2005; Ready, 2008). The case against the displacement hypothesis has been strongly stated by Clarke (1997b) and strengthened by Scott's (2004) extensive review showing that police crackdowns against crime can also be very effective and usually do not result in displacement. This is not to say that every crackdown or crime prevention effort is successful or that no crime is ever displaced, nor is it to deny that crime prevention must be resourceful. But it is a disgrace to use displacement as an excuse to hold back creativity in preventing crime.

The "Diffusion of Benefits"

Several criminologists are converging on what may be called a "diffusion of benefits" model (see Clarke, 1997b; Clarke & Weisburd, 1994). This states that crime prevention spills over. For example, closed-circuit television on one parking lot will tend to reduce auto theft, not only from that lot but also from other lots nearby.

This makes sense intuitively when we consider that fixing up your own house not only increases its value but also that of the house next door. For example, when a budget motel with problems of prostitution, drug use and sales, and loud parties is cleaned up and these crimes are reduced, most likely other types of crimes and problematic events will also reduce in the area around the hotel, such as car and pedestrian traffic issues, burglaries and robberies in the area, and loud noise complaints.

If one crime often begets another, one prevention effort should lead to another. The chain of relationships among crimes does not always fit one pattern, but in general, crime feeds off crime. That being the case, if we can learn how to reduce one type of crime, there is a chance that that will lead to additional good.

Crime as an Ecosystem

We have seen that reducing crime here does not automatically increase crime there. We also see that reducing crime here might reduce crime there. To bring these statements to life, one has to work out exactly when, where, and how. More generally, we have learned that crime is like an ecosystem. An ecosystem is a dynamic, living system of different activities, each drawing upon one another and on nonhuman resources (Felson, 2006).

A crime's ecosystem is its interplay with other crimes and the surrounding noncrime environment (routine activities). Illegal acts live off one another and off legal activities, just as foxes live off hares and hares live off crops. Offenders sometimes compete for turf, just as other life does. Crime suffers from destruction of its habitat, just as the deer population declines as cities take over and trees are cut down. Robbery grows with street prostitution, just as brown bears thrive with salmon to feed on. Like the ecosystem, crime has its dependencies and dynamics and must be studied accordingly (Cohen & Felson, 1979; Colquhoun, 1795/1969; Ekblom, 1997, 1999; Ekblom & Tilley, 2000; Farrell, 1998, 2001).

CONCLUSION

We often find that one criminal offense sets up a chain of criminal events. In most cases, heavy involvement in crime is a slippery slope for the individual, the victim, and the locality.[4] In the beginning of this chapter, we noted that the interplay of criminal acts can lead in either direction. On one hand, society suffers from a vicious spiral, with one crime leading to another, and then another. On the other hand, society benefits from an auspicious spiral, because prevention of one crime can easily thwart another. Removing crimes tends to set in motion a chain reaction of crime reductions going well beyond an immediate diffusion of benefits. Our best bet is to stop crime early, working on minor crimes, using discretion, common sense, and an understanding of how crime feeds crime. Our research agenda must be to learn a good deal more about these links and sequences. We also must learn everything we can about preventing crime early in its process. The next chapters give us some tools for accomplishing just that.

MAIN POINTS

1. Each minor offense can get the offender into deeper trouble right then and there. A good deal of minor offending is a slippery slope, sometimes leading toward violence and hard drugs.

2. Illegal markets are often interlinked with one another. For example, sex and drug markets, stolen goods and counterfeiting markets.

3. Repeat victimization is a very important part of crime. Fortunately, it also provides entry for effective crime prevention efforts.

4. A hot spot is an area or location containing much more than its share of crime or disorder events. Research has found that crimes and disorder events are disproportionately distributed across the geography of a city.

5. Crime facilitators are props contained in a setting, such as tools, weapons, drug paraphernalia, or disinhibitors, that help crimes along.

6. The concept of "broken windows" tells us that local deterioration and the proliferation of minor crime undermine a neighborhood, which may produce such a mess that major crime follows.

7. Zero-tolerance policing seeks to eliminate police discretion on both minor and major offenses; however, this is not realistic, and police must prioritize their enforcement activities.

8. Nibblers are those low-level offenders who nibble away at the public resource and make it less usable or bearable for everyone else. Grabbers are some urban offenders who seize for themselves large slices of what really belongs to the public.

9. Problem-oriented policing asserts that the goal of police should not be to arrest people or gain convictions but to reduce crime by addressing larger level problems in the community.

10. The Center for Problem-Oriented Policing (POP Center) is a virtual center that creates, warehouses, and distributes relevant research and practice on problem-oriented policing and, more generally, crime prevention strategies and research.

11. Civil and administrative laws are important to control settings that may facilitate crime, in that it is much easier to enforce civil and administrative laws against 30 owners than to try to enforce criminal codes against 10,000 citizens.

12. Displacement occurs when crimes move from one setting to another. However, research shows that a large number of crime prevention efforts

found that significant displacement seldom occurs. Diffusion of benefits is the idea that one prevention effort should lead to another.

13. A crime's ecosystem is its interplay with other crimes and the surrounding noncrime environment (routine activities). Illegal acts live off one another and off legal activities, offenders sometimes compete for turf, and crime suffers from destruction of its habitat.

PROJECTS AND CHALLENGES

Interview project. Interview one person about repeat property victimizations, trying to figure out why this occurred and what to do about it.

Media project. Search the media for coverage of an extreme case of college hazing. How did the activity evolve from minor to more extreme hazing incidents? Think about the groups that facilitate hazing rituals and why.

Map project. Map a student housing area. Pick out which student apartments make the best candidates for repeat burglary, noting them on the map.

Photo project. Photograph 10 possible indicators of decay in an area. Discuss exactly why and how each one might or might not lead to crime problems later.

Web project. Search for Web sites dealing with repeat victimization, reporting on what you learned.

NOTES

1. George Kelling and Catherine Coles elaborated on this in their book *Fixing Broken Windows* (1996). Although often countered (see Taylor, 2000), many practitioners and scholars have found the theory helpful. Consistent with their work, McGarrell and Weiss (1996) showed that improved traffic enforcement in Indianapolis reduced other types of crime. Joel Epstein (1997) has applied similar thinking to barrooms, and Ronald V. Clarke and Martha Smith (2000) have reviewed an extensive literature on preventing decline in public transit settings.

2. This is a "tragedy of the commons" (Hardin, 1968), except that it does not require global restrictions but rather simpler improvements in the design of urban space (see Chapter 9) or enforcement of simple statutes long in the books.

To learn about recent and dramatic changes in Chicago policing, see Skogan and Hartnett (1997) (see also Block & Skogan, 2001; Skogan, 2006; Skogan, Rosenbaum, & Hartnett, 2005; Skogan, Steiner, & Benitez, 2004; Skogan, Steiner, & Dubois, 2002).

3. One way to approach crime cycles is to consider very long-term changes in the lives of individuals. For example, scholars have correlated childhood experience with adult violent offending (Widom & Maxfield, 2001). Such studies are informative, but causation is difficult to establish clearly. Quicker shifts and shorter distances make hypotheses more falsifiable or verifiable. They also make it possible to figure out exact mechanisms leading from one crime to another.

NINE

LOCAL DESIGN AGAINST CRIME

———•◦•———

T his chapter is dedicated to five people, now deceased, who did a lot during their lives to help us understand and reduce crime: Jane Jacobs, C. Ray Jeffery, Oscar Newman, Barry Poyner, and Tim Crowe. All five made their contributions by showing us how to design cities and buildings to make them safer for people to live in.

On April 25, 2006, Jane Jacobs died at age 89. She wrote a classic book in 1961, *Death and Life of Great American Cities*. She was one of the first people to realize that the most secure neighborhoods are also very likeable and comfortable to live in. On the other hand, fortress construction often backfires by making it difficult to monitor the streets, making it *easier* for crime to occur.

Jane Jacobs has long been a hero to students of crime prevention because she showed how designing for less crime is as easy as designing for more. Her insights are still fresh today. She saw the coming tragedy of urban renewal before anyone else realized what was happening. She explained why old urban neighborhoods, even if low in income, provided places for pedestrians, had vibrant lives, maintained local control of space, and protected people against crime. These neighborhoods were bulldozed to erect unnatural high-rise public housing complexes that became sterile environments and had crime problems built into their design.

Her concept went beyond buildings themselves. It included the entire urban environment and took into consideration the people using that environment. Had people listened to Jacobs, U.S. housing policy would have made less of a mess than it did, but the errors go beyond housing policy. Overall, crime prevention strategies generally have been foolish and ineffective. Leadership fell into the hands of naïve ideologists, first liberal and then conservative. The first proposed to reduce crime by being good to people, thinking they would be good in return. The pendulum then moved toward an equally naïve position: Be bad to people to get them to be good.

Had either or both camps listened to Jacobs, they would have realized that the issue is not how good or bad the government is to people but whether public places are designed and organized to allow people to informally control their own environments. Even though we cannot turn back the clock, we can sometimes learn principles from the past and apply them today. Fortunately, we know much more than the barbed wire solution. It is possible to make buildings and homes warm and hospitable, while giving people cues and setting limits against crime (Felson, 1995a, 1995b). This chapter reviews some of what we know.

IMPORTANT IDEAS FOR DESIGNING OUT CRIME

C. Ray Jeffery coined the phrase "crime prevention through environmental design," which he used as the title of a book (Jeffery, 1971). The phrase was quickly shortened to CPTED (pronounced SEP-ted). Newman, Jeffery, and their followers generated important knowledge about the principles of lowering the risk of crime in local places and demonstrating these principles in real life. They helped clarify how to design new places more carefully so they did not foster crime, as well as how to fix old places to reduce their crime problems.

In 1972, an architect named Oscar Newman wrote a classic book, *Defensible Space*. His problem was how to design safer public housing.[1] His solution was not more locks or guards but rather a better design for how people use space. Newman divided local space into four categories according to degree of public access to it:

1. private,

2. semiprivate,

3. semipublic, and

4. public.

Newman's prescription was to move as much space as possible to the private end of the scale to increase security and prevent crime. He believed that people would look after their own private and perhaps semiprivate space, whereas people on the street would provide "natural surveillance" of semipublic areas. In pure public areas, surveillance of any type was difficult and crime risk would be greatest. Newman favored low-rise buildings over high-rise buildings. He also made something of symbolic divisions of space, such as low fences that were not true physical barriers but that clearly defined what was private. Newman supplemented his theoretical work and drawings with demonstration projects in the real world, showing major improvements in security.

Newman's suggestions were not always perfect. Paul and Patricia Brantingham (1975) point out that high-rise buildings for elderly populations and others having no children tend to be more safe and secure than low-rise buildings. Pat Mayhew and her associates (Mayhew, Clarke, Burrows, Hough, & Winchester, 1979) learned that natural surveillance by strangers is not really as effective as Newman had expected. That means that it is often necessary to assign someone clear and distinct responsibilities to look after public or semipublic space (a point made in the place managers discussion in Chapter 2). Symbolic marking of private and semiprivate areas does not necessarily work as well as more secure forms of access control. In short, we find that Newman teaches us a lot that is worthwhile, but we have learned more since his initial contributions.

Jeffery's CPTED provided additional information about which environmental designs are more secure. In addition, Jeffery trained, encouraged, or influenced several people who later made major contributions to the field of crime prevention. These excellent former students have done countless good in helping localities reduce crime and training others to do the same. They and others have helped to modernize CPTED.

ENVIRONMENTAL CRIMINOLOGY: A LARGER FIELD

Do not mistake environmental criminology for the study of toxic dumping: It is the study of how crime occurs in everyday life and how to prevent it. With this term, Paul and Patricia Brantingham (1984, 1998, 1999) have incorporated CPTED defensible space and similar practical concepts into a single perspective. Important concepts in this field are nodes, paths, and edges, as well as insiders and outsiders (refer to Chapter 2 for detailed discussion of these concepts, and to Chapter 3 for more on insiders and outsiders).

Space Patterns

How far does the offender go? What route does he take? Where are people victimized? What do offenders know about the places where they travel?

Questions like these are asked by "geographers of crime," and by many crime analysts. The Brantinghams incorporated these topics into environmental criminology. The nature of trips taken by offenders to crime sites is just one of its topics. Such ideas have been reviewed and extended in the work of Rengert (1996) and Harries (1999), and have influenced many parts of this book. Many police departments are deeply involved in mapping crimes throughout the locality. By learning more about how crime maps into the physical world, criminologists have done better in suggesting how to design more secure environments (Chainey & Ratcliffe, 2005; Eck et al., 2005).

Time and Space, Together

Environmental criminology considers space and time together. How do potential participants in crime move about through space and time in the course of a single day? Which hours of the day produce more of the routine activities leading to crime? Such questions come right out of the theory of *time geography* by Swedish luminary Torsten Hägerstrand (as chronicled by Pred, 1981). Hägerstrand showed how to trace each person's activity in the course of a single day. Psychologist Albert Bandura (1985) wrote about the significance of chance encounters, but Hägerstrand helped us to understand why such encounters are not as random as they seem. Tracing both your day and my day, we can see why we met somewhere or failed to do so. If we are on the same campus, have a class in the same building at the same time, and come from the same direction, we might see each other. What seems like a chance encounter in fact has a structure. The organization of time and space is central. It also helps explain how crime occurs and what to do about it.

Unfortunately, it is cumbersome to study time and space together. There is too much happening. To make sense of it all, we need theories to cut down on details and organize the ones we have.

The routine activity approach does that by finding three minimal elements of predatory crime: offender, target, and place. It joins environmental criminology in seeking to organize what we know about crime and everyday activities hour by hour, place by place. The routine activity approach also considers how hourly changes shift by days of the week, how daily patterns shift by month, how monthly patterns shift over the years, and so on (see Cohen & Felson, 1979; Felson, 1987; Felson & Cohen, 1980). It finds dramatic shifts in everyday life, helping to explain crime rate trends. This is very much in agreement with the landmark research by Hindelang et al. (1978), which emphasized how lifestyles contributed to crime rates. These early lifestyle and routine activity studies, however, did not develop ideas for preventing crime.

Putting Knowledge Into Practice

Rather than doing science separately from practice, environmental criminology went onward to try preventing crime. The Brantinghams developed their thinking and applied their work in Vancouver, a large city in the western Canadian province of British Columbia. This city has the size, climate, and "feel" of Seattle, Washington. Forging relationships with local architects, planning boards, and police, the Brantinghams helped other people apply environmental criminology. The results include

- A public housing complex with very moderate crime rates, looking clean and neat
- A new town with a recreation center that avoids fights, despite a large influx of male strangers
- A winter sports village with fewer problems from drunken revelry once activities were somewhat reoriented
- Local business areas with less lunchtime shoplifting, the result of school lunches provided by local churches

One of the former students of the Brantinghams is Mary Beth Rondeau, a former Canadian Olympic swimmer, trained architect, and staff associate of the planning department for the City of Vancouver, British Columbia (see Vancouver Planning Department, 1999, for excellent photographs and drawings of these accomplishments). Rondeau has helped ensure that a good deal of the new construction in Vancouver minimizes crime opportunities. New townhouses and many other buildings in Vancouver look right at the street and send a clear cue: Don't even try it. Step by step, that city is slicing away at its crime environments. Similar progress is showing up in Newark, New Jersey, where new housing displays common sense, informed by preventive thinking.

One of the most important applications of environmental criminology in western Canada was carried out by the Royal Canadian Mounted Police, which began to train its officers in crime prevention. These Mounties are trained to read blueprints; they routinely sit on planning boards, examining detailed plans for new construction and making suggestions to help reduce the risk of crime. As a result, many new housing complexes, businesses, and schools in British Columbia apply principles of environmental criminology from the outset. Similar training of law enforcement personnel is also found in Britain and the Netherlands, where reducing the opportunity for crime is a significant part of government policy.

Tim Crowe (1991) did more than anybody to teach local governments and school administrators in the United States and around the world how to design out crime. Devoting his life to that goal, he died in 2009, while we were working on this chapter. Tim's favorite remark was that "good design is safe design." This means that crime can usually be reduced *without* unpleasant trade-offs. Indeed, comfortable places are usually safe places. Most people have a kind of sixth sense about a place, telling them whether it is safe and comfortable. They are not always right, but environmental criminology confirms common sense more often than not. It also helps refine the scientific principles by making sure that they fit reality.

Another great designer against crime is Diane Zahm at Virginia Polytechnic University. She has applied her considerable experience to employing the urban and metropolitan planning process to reduce crime in many nations and many states within the United States. The results of these planning efforts are incremental: one town at a time, one building here, one plaza there, an improvement to a part of the college campus, a safer transit system, and so on. A number of urban scholars and metropolitan planners have employed terms such as "livable communities" or "livable cities" to describe an approach to design that encourages pedestrians, "feels" better, and also is more secure from crime.

The Police Role

Those trained in environmental criminology cannot always prevent crime. Sometimes, though, environmental criminology can help police to detect and thwart offenders who already have gotten started. An excellent illustration is the effort to stop serial rapists and serial murderers. Kim Rossmo is a leader in mapping out these criminal acts and deducing a good deal about these offenders. Rossmo (1995, 2000) studies exact times and locations of each serial offense and follow-up events—for example, serial rapist-murderers find the victim in one place, perform the sexual attack in another, and dump the body in a third location. He puts environmental criminology and theory to work to figure out where the offender most likely lives, works, and has his recreation (see also Beauregard, Proulx, & Rossmo, 2007; Rossmo, Davies, & Patrick, 2004). This information aids in figuring out the identity of the offender and making an arrest. The more interesting point is that problem-oriented policing increasingly finds common ground with environmental criminology. More recently, Rossmo and colleagues have shown the larger scientific community that shark attacks fit patterns similar to human serial offenders. That is, sharks forage for food much as offenders search for their victims.

As we study offender behavior in space, we learn that certain places are very likely to contribute to crime. Such places often have legitimate purposes as well. For example, fast-food restaurants serve food to millions of people, but also provide locations for offenders to meet. The location of fast-food restaurants that cater to teenagers has an extra impact on local crime problems. Solving these problems may tell us to look closely at the location decisions for such restaurants. The Brantinghams (Brantingham & Brantingham, 1984) have shown that proximity to a McDonald's restaurant is an excellent predictor of nearby property crime as well as other offenses.

Mangai Natarajan (2000) examines transcripts from police wiretaps of conversations among those dealing large amounts of drugs. These offenders apparently like to meet in fast-food restaurants to make their deals. Cooperation with business to prevent the use of their premises for criminal behavior is an important theme of Felson and Clarke (1997a) and of problem-oriented policing. Design features can be important. There has been at least one McDonald's restaurant in Manhattan that found out retail drug dealers had been cutting the cushions to insert packets; as a result, they installed a new type of chair without cushions to make their environment less favorable to crime.

Basic Principles

In recent years, experts in crime prevention have become more eclectic, drawing ideas from "defensible space," CPTED, and principles of good management. In an excellent essay about how to design out crime in local areas, Tim Crowe and Diane Zahm (1994) state three design approaches to reducing crime:

1. Control natural access.

2. Provide natural surveillance.

3. Foster territorial behavior.

Controlling natural access does not require huge locks or walls. It includes hedges and shrubs, gates, doors, and plans for walkways—anything else that encourages people to go where they will do no harm or receive no harm. An example is a hospital with hundreds of people visiting patients each day, many of them entering the wrong door. The true visitors kept getting in the way of surgery, and some people entered there to steal things. The hospital just locked that door and removed its handle, attaching a sign to enter through the visitors' door, with an arrow in that direction. The problem was solved simply, without requiring arrests or punishment. Providing natural surveillance often is achieved simply by trimming hedges. Fostering territorial behavior often is accomplished by putting in small fences or porches, or landscaping to show where one apartment ends and the other begins. The three design approaches above apply the three strategies below:

1. *Natural strategies:* Security results from the design and layout of space. Both human and capital costs are low.

2. *Organized strategies:* Security guards or police play the central role (George & Button, 2000). These strategies are labor intensive and expensive.

3. *Mechanical strategies:* Alarms, cameras, and other hardware are employed to control access and provide surveillance. This may require additional employees to watch monitors or respond to alarms. The equipment may be expensive.

Clearly, natural strategies are superior in economic terms, and they also avoid confrontation by preventing crime from happening in the first place.

From Newman, Jeffery, and subsequent experience, we learn that natural strategies work best. It is indeed mysterious that so many people think otherwise. Widespread ignorance leads too many commentators to say, "Oh, you mean locks, alarms, and guards." In fact, the main idea of designing out crime, rightly understood, is to achieve safe environments in far more intelligent and less costly ways. That means designing a neighborhood with the human scale foremost, fostering communication among people, and encouraging their sense of ownership and responsibility. This contrasts with the stark methods of developing no-man's-lands, walls, and thick barriers. Modern crime prevention uses a broader repertoire rather than always playing the same tune. That does not rule out the use of fences in the right places, but it includes much more, as we shall see.

Crowe and Zahm sum up much of what we know about designing out crime with two words: *marking transitions*. People should know when they are entering your space, and you should know when you are entering theirs. Marking transitions provides reminders. Someone trained to design out crime will provide clear markings for and separation of

- Controlled and uncontrolled space
- Public, semipublic, semiprivate, and private space
- Conflicting activities

The interesting point here is that completely legal activities can produce illegal outcomes. A good thought experiment is to find and discuss what activities conflict in a way that invites trouble. Consider

- Teenagers with stereos versus elderly people who want quiet
- Defecating dogs versus park visitors sitting on the grass
- A hangout for older kids versus a playground for toddlers
- The high school versus the shopping mall (see Chapter 6)
- A bikers' bar next to a gay bar

Designers can find creative ways to use hedges or walkways to separate conflicting activities. Businesspeople, too, can pay more attention to location decisions.

We often think of CPTED in spatial terms, forgetting its substantial temporal aspect. Scheduling and coordination of activities in time and space play an important role in assembling likely participants in crime. Some schedules produce stragglers who are easy to attack. To prevent that from happening, it makes good sense to

- Schedule activities carefully in time and space for more effective and intense use, and hence less chance for stragglers to be attacked or for groups of likely offenders and suitable targets to assemble
- Design pedestrian paths that intensify usage for greater safety

We have to admit that some places, as well as some human activities, are implicitly risky from a crime viewpoint. So what should be done? If we put the unsafe activities in unsafe places, a lot of trouble probably will emerge. Crowe and Zahm (1994) have made this remarkable and insightful recommendation for everyday design:

The Crowe-Zahm Mixing Principle: place safe activities in unsafe locations and unsafe activities in safe locations.

The logic of this principle is to use safe locations to calm, contain, or help supervise activities that would otherwise be dangerous and to use safe activities to provide natural surveillance of potentially dangerous places. Our team of graduate students studying the subway stations in Newark, New Jersey, recommended putting a small business beside each of the riskier subway stations to provide more eyes on the street (Felson et al., 1990). We cannot make all of life safe, but we can minimize the dangers and risks by following the Crowe-Zahm advice. Why give offenders their best shot at crime?

Physical Aspects of Crime Prevention

Crime can be prevented by at least four physical methods: target hardening, construction, strength in numbers, and noise. For example, universities harden targets when they bolt down computers, typewriters, television sets, projection equipment, and the like. As for construction, universities sometimes put up extra walls, fences, or other physical barriers to reduce unauthorized entry to university buildings, or simply to channel flows of people coming and going, which makes mild supervision possible. The University of Southern California, for example, put a fence around the premises in preparation for the 1984 Olympic Games. The campus was still open at many gates, but offenders could no longer enter anywhere, attack anything or anyone, and exit anywhere afterward.

Strength in numbers is important for helping people to protect themselves and their property. One designer of a high-rise building for the elderly put the recreation room on the first floor with good lines of sight to the door. Together, the residents were able to keep people from wandering in without permission.

Noise also is important for crime. On the negative side, offenders use noise to determine whether you are present or absent, even banging on your door to make sure you are gone. On the positive side, noise can protect you. A good lock on your door is important not so much for preventing illegal entry as for making sure the offender makes enough noise to draw the attention of others. Alarm systems operate on the same principle. Noisy dogs can serve to both alert others and scare off offenders directly. Noise also can be directed at offenders, as when subway station personnel with loudspeakers direct someone to cease an undesirable behavior. In either direction, noise reminds us that crime is a physical act and that our five senses are essential for both committing it and preventing it.

Lines and Routes

Those who work on designing out crime pay a good deal of attention to sight lines, that is, the visual lines from a potential crime target or entry point to potential guardians. Good sight lines generally thwart criminal activity if someone is around to discourage it. But sight is not the only issue. Paths for entry and exit also influence how potential offenders and victims calculate their chances. The Brantinghams (1984, 1998, 1999) pay a great deal of attention to paths people take in daily life and the risks incurred or crime chances provided. Fisher and Nasar (1995) consider how routes and their enclosure influence the offender's ability to get to a victim, or the victim's ability to evade an offender. The challenge is to design clear sight without much danger of illicit entry and exit. Nowhere is this more relevant than in housing.

RESIDENTIAL CRIME PREVENTION

On November 26, 2006, Barry Poyner died. He was a most unusual architect who developed a career figuring out how to improve the lives of ordinary people by designing safer buildings and roads (Webb, 2007). He brought these insights to the newly emerging field of "design against crime"—a phrase he popularized. He developed new methods to make housing more secure, and his guidelines were applied to housing across Great Britain (Poyner, 1998).

For example, Poyner paid great attention to whether residents have a clear view of the street outside or whether their view is blocked. His recommendations include ways of reducing four different types of crime through better residential design, including residential burglary and vandalism; car crime near home; and thefts in the front of the house, in yards, gardens, garages, and sheds.

The more we learn about residential crime prevention, the more we realize that it applies at all income levels. Much of the original work considered reducing crime in public housing or other residences for the poor. Poyner showed that these principles also apply to middle-class housing, and Rengert and colleagues found that more secure designs also work well in higher income areas (see also Hakim, Rengert, & Shachmurove, 2001; Rengert, Mattson, & Henderson, 2001; Rengert & Wasilchick, 2000). The same principles also apply in very different nations and over many locations within a single nation (Edmunds, Hough, & Urquia, 1996).

Poyner and Webb (1991; Poyner, 1998) also learned that offenders on foot operate differently from those in cars, requiring different prevention measures. They offered seven recommendations for designing the house and property itself:

1. Use proper design so houses only need moderate locks.

2. Make front windows face those across the street.

3. Fence the back and side yards.

4. Position front service and delivery areas carefully.

5. Leave a garden area in front of the front fence.

6. If there is an open car park area, put it in front of the house.

7. If there is a garage, put it at the side, near the front entrance.

The front windows that face across the street allow neighbors to supervise the front area and to notice anybody breaking in or stealing something from the front. Exhibit 9.1 shows how much easier it is to see what is going on when the street is designed that way. Secure back and side yards make it more difficult to break in there. Fenced and gated fronts also can thwart theft and break-ins, especially if they can be seen from inside the house, if the gates are bolted from the inside, and if the service and delivery areas are easy to see from everywhere.

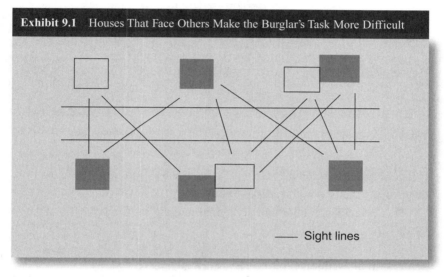

Exhibit 9.1 Houses That Face Others Make the Burglar's Task More Difficult

—— Sight lines

SOURCE: Adapted from Poyner, B., and Webb, B. (1991). *Crime Free Housing* (p. 98). Oxford, UK: Butterworth.

How high should the fence be in back and on the side? In Britain, houses are often smaller than in America, so a tall fence or wall may do the job, but large American yards make it all too easy to break in without being heard. We would shy away from thick or tall fences or walls in that case and make sure the fences do not block the sight of neighbors or vice versa. The garden space of about 10 feet in the front, if not too bushy, keeps people away from the fence, guiding them through the gate for legitimate business.

Exhibit 9.2 describes a small house that is fairly secure, with some fencing in the back, on the side, and in front of the house, and a small front garden to set the fence away from the street. Finally, if parking is outside, you should be able to look out the front and see that your car is still there. An inside

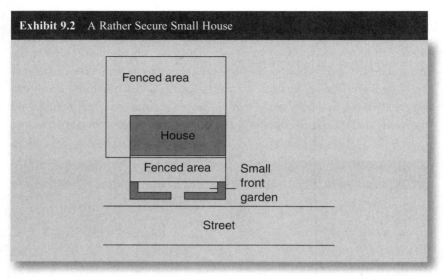

Exhibit 9.2   A Rather Secure Small House

SOURCE: Adapted from Poyner, B., and Webb, B. (1991). *Crime Free Housing* (p. 99). Oxford, UK: Butterworth.

private garage can be on the side, but its entry should not be far from the front door, so any thefts would be easy to notice.

Poyner and Webb offer four additional suggestions for design of an entire locality:

1. Minimize through traffic on residential streets.

2. Make sure pedestrian paths have dead ends.

3. Orient fronts of houses toward points of entry.

4. Keep any park areas outside the housing areas.

Separate pedestrian paths are common in Britain and American public housing projects. These paths are often used by offenders to find their targets and then to exit with the loot. When pedestrian paths are constructed without a through route, meeting a dead end within the residential area, they no longer encourage household crime. A pedestrian will not be able to walk by your place, take what he wants, and continue in the same direction. In general, the best design probably is to keep pedestrians walking on sidewalks in front of houses and beside streets, with each household opening onto the street.

Public housing complexes that have followed this rule have had relatively few crime problems. Public housing with internal through paths and apartments opening onto a central area have far more difficulty with crime. Even worse, some public housing suffered greatly from putting in overhead walkways linking buildings, making it easy to commit crime and escape to the next building. When these walkways were closed off, the crime problems went down right away (Poyner, 1997).

Parks inside housing areas look very good in architects' models before a place is built. After the housing is constructed, however, parks can easily become a mess, welcoming the toughest and rowdiest youths and even being taken over by gangs. Green spaces provided *just outside* a residential area rather than within it produce less crime. The challenge to those who love parks is clear: Find a way to design and locate park areas so they do not become *easy routes for illegal entry or ideal for illicit takeover.*

The human race has never found a better urban orientation than square or rectangular houses lined up on a street and opening to the front (Exhibit 9.1). If you want to go to your neighbor's house, you should go out your front door, walk down the sidewalk, up to your neighbor's front door, and knock. Yet planning has to be flexible. In a quiet cul-de-sac, it is possible to orient houses in the direction of the busy street so residents can notice who is entering. Exhibit 9.3 depicts this arrangement and draws the sight lines. To figure out whether and when this will succeed, think about what direction the traffic is moving and from where an offender is likely to arrive.

Street Closures and Alternatives

In general, through streets with lots of traffic generate a good deal of crime risk, but many existing environments already have through streets. Sometimes, streets can be closed off to reduce crime. Ron Clarke (2005, pp. 5–6) outlines some benefits of closing streets or alleys:

- Criminal outsiders are less likely to become familiar with the area.
- Residents learn who does not belong in the neighborhood, which helps them to more effectively keep watch on the streets near their homes.
- Residents committing crime in their own neighborhood cannot easily blame outsiders and thus deflect suspicion from themselves.

Exhibit 9.3   Orient So Those Entering From the Larger Street Are Noticed

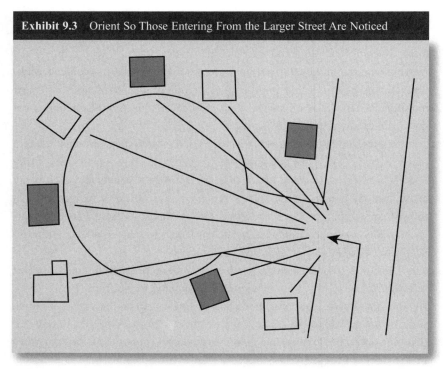

SOURCE: Adapted from Poyner, B., and Webb, B. (1991). *Crime Free Housing* (p. 101). Oxford, UK: Butterworth.

- Burglars cannot easily gain access to properties, especially from alleys behind houses.
- Escape routes for robbers are blocked off.
- Drive-by shootings are prevented because cars cannot easily enter a street, or because they have to backtrack to escape, exposing them to retaliation from those shot at.

The idea of privatizing streets was noticed in St. Louis by Oscar Newman, who then popularized the idea. Gardiner's (1978) book *Design for Safe Neighborhoods* showed how he redesigned Asylum Hill in Hartford, Connecticut, closing off some streets to cars without interfering with pedestrian access. Crime went down immediately. It is difficult to argue that these crime reductions were purely coincidental.

More recently, Los Angeles has experimented with creating some cul-de-sacs in a very dangerous local area with lots of drive-by shootings. When streets were blocked off without much thought, they achieved relatively little. When some streets were closed to cars on a strategic basis—calming traffic but still allowing a natural flow—drive-by shootings declined (Lasley, 1998). This points to a general lesson of designing out crime: Be attuned to local facts and realities.

Critics of closing streets often say this is a method for excluding people, but to reduce crime, street redesign does not have to exclude people at all. Exhibit 9.4 displays some simple ways to slow down traffic and reduce residential crime risk

Exhibit 9.4 You Do Not Have to Close Off a Street to Reduce Crime—Just Calm the Traffic

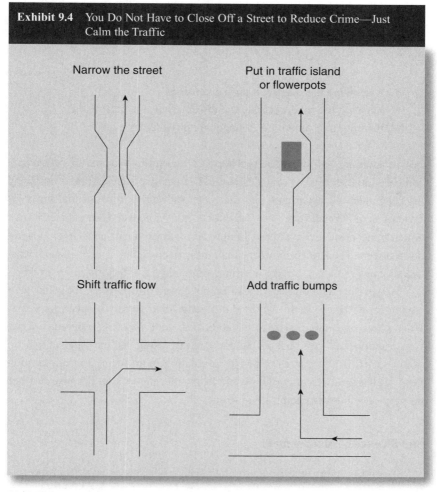

SOURCE: Adapted from ideas borrowed from Patricia and Paul Brantingham, personal conversations.

without closing off streets. Traffic engineers call these techniques "traffic calming." People often have vast differences of opinion about traffic calming. Younger people who have no children are on the go and usually do not want to slow down. Older people who have young children like the idea, especially because the slower traffic is less likely to run over or scare their children. This alerts us to the complexity of crime prevention in practice and the many choices we must make. It also reveals the importance of many alternatives, just in case one solution leads to another problem that's worse.

Whatever your opinion, traffic calming techniques have been shown by research to be particularly effective in reducing speeding in residential areas (Scott, 2001) and are a way to reduce some types of crime at very little monetary cost. According to Scott (2001, pp. 10–11), common methods that affect the psychology of speeding are as follows:

- Narrowing the road with curb extensions
- Marking the road to create the illusion that it is narrowing
- Planting trees and other foliage along roadsides

In addition, traffic circles, roundabouts, and traffic islands for pedestrians compel motorists to slow down. Some traffic engineers have slowed traffic by inserting sidewalk extensions into the road, gateways into residential areas, or dividing up a through street into two cul-de-sacs. Moreover, road engineers have gotten quite creative with speed bumps and humps, small and large, slowing down drivers. The British sometimes call these sleeping policemen—a good term because they require no salary, overtime, or training, and they make no arrests.

However much you may dislike being slowed down, it is a lot better than having crime take over in the area and better than getting a speeding ticket. Hilly cities and suburbs offer an additional tool for designing out crime. Construction has to adjust to the topography, using inclines and curves in doing so. Hills and curves can help in creating more local control, calming down traffic, and discouraging people from entering with no purpose at all or no purpose other than committing crime.

Significance of Management

Despite its very bad image, public housing varies greatly in crime risk (Holzman, 1996; Hope, 2007; Weisel, 2002). Eck's (1995) pathbreaking work on how drug sales find their niche within public housing or private low-income housing shows that poorly managed housing often loses some benefits of good design features, whereas good management can offset bad design.

An active place manager, such as an apartment supervisor, is essential. Eck found that drug sales were made within those apartment buildings lacking a manager, or by a manager who was corrupted by the drug dealers (Eck, 1995). Management goes beyond a single place manager. Put yourself in this person's position: If the larger organization gives you no rewards, no security, and no backing, or even punishes you for taking the initiative, why should you stick your neck out to fight crime? Good management has to work at many levels, but it can make things better than they were. A doorman, concierge, or janitor can make a great difference in protecting any building.

OTHER METHODS FOR DESIGNING OUT CRIME

In their book *Design for Inherent Security* (1995), Barry Poyner and W. H. Fawcett suggest ways to design out crime in nonresidential buildings. They show how to design low crime risk into shops and stores, malls, offices, industrial parks, hotels, sports buildings, bars and restaurants, transportation hubs and stations, public parking garages, service stations, hospitals, schools, and more. We do not have the space to cover all these applications, but we can examine some.

Convenience Stores and Small Groceries

Small stores with late hours are very vulnerable to robbery, especially when located near freeways. In some cases, employees or customers have been injured or murdered in the process.

One large convenience store chain, 7-Eleven, suffered staggering increases in the number of robberies during the 1970s. The chain's owner, the Southland Corporation, hired the Western Behavioral Sciences Institute and several convicted robbers, including Ray Johnson, to help redesign the corporation's retail stores with the intention of reducing the number of holdups. Sixty stores initiated the team's recommendations, and a control group of the same size did not. The control group experienced no change in robbery risk, whereas the experimental group had a 30% decline in the number of robberies.

These stores also reported major declines in nearby crime and in people loi-tering and harassing customers. These are some of the innovations carried out (see Duffala, 1976; see also Altizio & York, 2007; Matthews, 2002; Mayhew, 2000):

- Do not let display advertising cover the windows and protect robbers from being seen from the street.
- Move cash registers to the front of the store, where they are visible from the street.
- Put cash registers and clerks on a raised platform, taking the cash drawer out of the offender's line of vision and making the clerk more imposing.
- Install timed-access drop safes beneath each register, releasing no more than $10 change every 2 minutes. This removes the target of the crime.
- Redesign store properties to eliminate all alley exits, channeling every-one through the front.
- Encourage taxis to use the premises as a nighttime station, giving drivers free coffee and restroom privileges.
- Train employees to make eye contact with each customer on entry.

Most of these methods produce more natural social control. Although some of these policies go beyond CPTED and defensible space, they show that it is possible to make stores safer with simple place management strategies and without using armed guards or making them less comfortable environments. At the urging of Professor C. Ray Jeffery, the city of Gainesville, Florida, placed similar stipulations into a city ordinance and also required the presence of at least two clerks on duty late at night. The result was a decline in conve-nience store robberies (see Hunter & Jeffery, 1997).

In general, a well-designed and safer store not only reduces crime but also draws more customers, who in turn enjoy shopping there. No better illustration of this principle can be found than in the next example.

Design by Disney

Clifford Shearing and Phillip Stenning (1997) have reported how Disney World organizes visits in great detail. From parking lot to train to monorail to park and back, activities are planned to minimize risk of accident or crime. Disney World follows this rule: Embed control in other structures so that it is barely noticed. For example, entry into most exhibits occurs only within a vehicle controlled by Disney personnel. Even lines with people waiting are wrapped back and forth by rails, to encourage informal conversations, mixing of adults and children, thus reducing impatience and discouraging rowdiness. (In more recent years, Disney has done away with many lines, allowing people to register their places electronically and thus not have to wait until their turn comes up.)

The most important lesson of Disney planning is that almost all visitors to Disney World are quite contented with the way their visit is managed and comply voluntarily. Of course, people are not spending the rest of their lives in Disney World. We learn that crime prevention can be most effective when it is incidental and that a well-planned and well-managed environment serves many human purposes along with security.

A Huge Bus Terminal Turns the Corner

You can learn a good deal about designing out crime from the substantial changes in New York City's sprawling Port Authority Bus Terminal (Felson et al., 1996). This building contained a good deal of crime within and had evolved into a messy place to control. It was impossible to reconstruct the building to incorporate what we now know about designing out crime. There were already more than 100 police officers assigned to the building, and they could not hope to control the place. Management hired various consultants, including a group called the Project for Public Spaces, which specializes in making public areas livable and usable.

In the early 1990s, major changes were made to regain this building for its intended purposes. To accomplish the task, it was necessary to gain control of the transients who stayed in the bus terminal day and night. Courts had recently ruled that people did not have a constitutional right to panhandle or linger in the terminal if they were not using it for transportation. Transportation agencies gained additional standing in law and public relations by offering programs designed for these transients. The Port Authority created Operation Alternative, which offered transient people a good variety of social services. The other alternatives were to leave the building or go to jail. This is one of those cases in which the kinder and gentler methods were assisted by the tougher and meaner ones.

Without knowing it, the Port Authority also applied the Crowe-Zahm Mixing Principle. It put safe activities in unsafe locations, moving pushcarts selling flowers into abandoned corners. It put unsafe activities into safe locations, concentrating ticket purchases in a convenient but not cluttered area, so people taking their wallets out would not have them snatched.

The bus terminal made many small but noteworthy physical modifications:

- It designed entrances and escalators to flow better and to move crowds of passengers through more quickly. This made it harder for illicit activities to keep control. Open entries replaced the dreary and dangerous ones.

The Port Authority

- *Removed niches and corners.* It filled or closed off empty spaces so no one could hang out there getting drunk or high. It narrowed or connected columns to deny illicit activities a place to hide or linger.
- *Improved bus gates.* That prevented people from living within or above them.
- *Controlled emergency staircases and fire doors.* This denied places for sleeping, injecting heroin, or other violations.
- *Reduced the seating.* With fewer places to hang out and less comfortable seats, long-term lounging was not easy in a facility supposed to take care of quick transit.
- *Cleaned and shined the floors.* This, plus ongoing maintenance and better lighting, kept the place bright and less fearsome.
- *Removed the seedy stores.* These were replaced with upscale chain stores that commuters would like.
- *Designed away sex pickups.* A new control room between the balcony and first floor blocked the view of those in search of partners for quick sex, ending the "meat rack."

Restrooms were a particular problem in the bus terminal. They were sites for homosexual liaisons, thefts of luggage and wallets, and drug taking. Homeless people removed ceiling tiles and moved into the ceiling area, or just hung out, taking full baths in the large sinks and monopolizing facilities. There was no room for legitimate relief. To solve the problem, the Port Authority renovated the washrooms (toilets) dramatically, as summed up in Exhibit 9.5. They removed ceiling panels and obstructions, designed stall doors so the number of legs could be counted, and designed in cleanliness and light. The rooms were enlarged and placed in areas with merchants near and natural surveillance.

Exhibit 9.5 Making Washrooms That Discourage Crime, Port Authority Bus Terminal, New York City

Washroom (Toilet) Characteristics	Before Changes	After Changes
1. Ceiling panels	Removable	Secure
2. Stall doors	Tall, low to ground	Less so
3. Stall walls	Easy to write on	Resistant to writing
4. Ventilation	Poor	Good
5. Corner mirrors	Absent	Present
6. Sink size	Six users each	One user each
7. Fixture controls	By hand	Automatic
8. Lighting	Poor	Good, secure
9. Tile	Small squares, dark	Big, bright tiles
10. Walls	Angled	Straight
11. Nooks	Present	Absent
12. Stores	Far from entry	Near entry
13. Size	Small	Large
14. Attendants	Absent	Present

SOURCE: Data from Felson, M., Belanger, M. E., Bichler, G. M., Bruzinski, C. D., Campbell, G. S., Fried, C. L., et al. (1996). Redesigning Hell: Preventing Crime and Disorder at the Port Authority Bus Terminal. In R. V. Clarke (Ed.), *Preventing Mass Transit Crime*. Monsey, NY: Criminal Justice Press.

NOTE: The changes were made to the washrooms in the early 1990s.

Attendants were provided, the only change that required higher labor costs in the long run. Most important, the Port Authority provided much better management of its facilities, reinforcing the improved design, offering better services for the homeless, and generally keeping the flow of legitimate activities going.

These changes were accompanied by important reductions in the pornography business in the Times Square area. Although the bus terminal changes moved along somewhat faster, they reinforced each other. Robbery and assault rates went down dramatically inside the bus terminal as well as in nearby police precincts. Even New York City saw improvements in its crime rates (see Kelling & Coles, 1996). The bus terminal did better. For example, with 1991 indexed at 100, robbery rates went down to 66 in 1992, 43 in 1993, and 30 in 1994. That is a 70% decline. The police data reveal serious declines in major and minor complaints.

The Port Authority showed a lot of ingenuity in doing public surveys to see whether customers noticed its efforts. Research staff rode random buses, passing out questionnaires to customers. Seated with nothing much else to do, most people filled out the form. Over the years, these forms showed major public recognition of greater cleanliness, fewer hassles with panhandlers and drunks, and a feeling of safety.

In the bus terminal, the Port Authority also has maintained nightly checks of the number of homeless people hanging out. In 1991, researchers counted an average of 151 homeless people per night. In 1992, this was down to 124. In 1993, it declined to 60, and in 1994, it was down to 30. Meanwhile, the numbers involved in the homeless programs sponsored by the Port Authority rose significantly.

Designing a Low-Crime Subway System

Nancy LaVigne (1996, 1997) offers a remarkable account of the design of the Washington, D.C., subway system. That system is almost the opposite of the old systems in America. Its stations are voluminous and have excellent "sight lines." In other words, it is difficult for an offender to act or escape unseen. Entries are visible, and pedestrians are funneled into spaces where they can provide mutual informal protection. The net effect is a remarkably low crime rate, even though the subway is located within a high-crime city. The main crime problems are in the parking lots outside the suburban stations, where cars are targets of crime.

Designing Safer College and
University Campuses and Schools

Many American college and university campuses have a lot of crime. Cars, their contents, or accessories are stolen. Wallets are removed from offices, libraries, and dorms. Electronic equipment is taken. Sometimes, personal attacks occur. Important research on college campus victimization by Bonnie Fisher and her colleagues at the University of Cincinnati (Fisher, 1999; Fisher, Cullen, & Turner, 2000; Fisher, Daigle, & Cullen, 2003, 2007; Fisher & Nasar, 1995; Fisher & Sloan, 1995; Fisher, Sloan, Cullen, & Lu, 1997) helps confirm the routine activity aspects of student and staff victimization and brings to mind the possibilities for designing out crime (Brantingham & Brantingham, 1994, 1998, 1999, 2003; O'Kane, Fisher, & Green, 1994; Zahm, 2004; Zahm & Perrin, 1992).

Campus parking areas are especially subject to risk of property and even violent victimization. Here are some ways campus parking areas can be made more secure:

- Arrange for nighttime students and workers to have parking near building doors, but not so close that they block the view of the parking area from the building.
- At low-use times, close off unneeded parking areas or sections of large parking areas to concentrate cars and people for supervision.
- Require students and staff to sign up by name and have a sticker, even for nighttime or free areas.
- Get visitors in cars to sign in and give them time limits.
- Fence parking areas.
- Eliminate nooks and corners in parking structures.
- Build parking structures as slopes so people on foot will have clear sight lines.
- Make parking structure stairwells easy to see into.
- Orient buildings to face parking areas.
- Trim hedges and the lower limbs of trees around parking areas, and avoid thick foliage.
- Post signs and organize the flow of traffic so neither cars nor pedestrians will get lost.

The list includes cheaper and simpler items, such as trimming hedges, closing off unused parts of the parking areas, and focusing students and staff into a more limited area. It also includes ideas that would apply only to new construction, such as orienting buildings so people can see parking lots and sloping the parking structures for natural surveillance inside.

Now that you have read about CPTED, have a look back at Chapter 6. You will see that many of the ideas presented there fit well within the current chapter. More compact schools and school grounds make it much easier to supervise against crime. Smaller student enrollments per school can be interpreted along the same lines. Locating schools for greater security is also a design concept. In addition, some states pay close attention to specific design principles when constructing high schools (see Fein, Vossekuil, & Pollack, 2002; Kuenstle, Clark, & Schneider, 2003; Sprague & Walker, 2004). As you progressed in reading this book, you probably noticed that the ideas in one chapter largely fit with those in another. You can see that crime is highly local, that it can be prevented locally, that it responds to physical features of everyday life, that these features provide cues to likely offenders, and that environmental design influences what cues are emitted in the course of daily life.

CONCLUSION

Many people have developed solid principles for designing out crime in local places. Such principles lead to specific advice and ideas that are accessible to people in every walk of life. With these principles, crime can be sliced at here and there with a good deal of net success. In meta-analyses of professional evaluations for various types of crime prevention, these methods fared far better than others (Clarke & Eck, 2007; Eck, 1997; Weisburd et al., 2004).

Society does not have to wait for trouble and pick up the pieces later. Designing out crime offers a kind of natural crime prevention, including (a) unplanned and informal crime prevention as it occurs naturally in everyday life and (b) planned crime prevention that imitates the former by skillfully designing human settings or activities so that crime opportunities are reduced unobtrusively and nonviolently. A central theme of crime prevention in this chapter and throughout this book is that crime prevention works more naturally when human activities are divided into smaller and more manageable chunks. These chunks can help provide social control and thus crime control. Responsibility to stop burglary can be linked to those in the immediate area.

A school can be situated for more local visibility, and public housing can be designed or sized for better control. High-rise buildings can be arranged so that elderly people can protect their piece of the world. A shopping area, too, can be localized for more control. Activities also can be channeled for more informal and natural social control. Thus, chunking and channeling are the main tools of designing lower crime rates into everyday life.

Criminology in the past has swung between utopianism and hopelessness. We are arguing for something different: a nonutopian optimism. We can reduce crime here and there but not forever. We can make major achievements but have to keep striving. Designing and managing local places to reduce crime is an important part of the formula; crime prevention goes beyond that, however, as the next chapter shows.

MAIN POINTS

1. Designing buildings and environments to prevent crime need not create a fortress. Try to control natural access, provide natural surveillance, foster territorial behavior, or create some combination of these ideas. In general, make the space hospitable for legal activities and crime usually will go down significantly.

2. The three design approaches in the previous point apply these three strategies: (a) *natural strategies:* security results from the design and layout of space; (b) *organized strategies:* security guards or police play the central role; and (c) *mechanical strategies:* alarms, cameras, and other hardware are employed to control access and provide surveillance. The latter two are often expensive, so natural strategies are superior in cost, and they also avoid confrontation by preventing crime from happening in the first place.

3. The Crowe-Zahm Mixing Principle: place safe activities in unsafe locations and unsafe activities in safe locations. As places feel better for people, they also become more secure.

4. Crime prevention through environmental design (CPTED) is made up of principles that seek to lower the risk of crime in local places by design and dictate the use of the physical environment.

5. Environmental criminology is the study of how crime occurs in everyday life and how to prevent it.

6. Understanding how crime happens at certain places and at certain times is central to explaining how crime occurs and what to do about it.

7. Crime can be prevented by at least four physical methods: target hardening, construction, strength in numbers, and noise.

8. One of the best applications for designing out crime is housing. Seven recommendations for designing a house and property itself are as follows: (a) Use proper design so that houses need only moderate locks; (b) make front windows face those across the street; (c) fence the back and side yards; (d) position front service and delivery areas carefully; (e) leave a garden area in front of the front fence; (f) if there is an open car park area, put it in front of the house; and (g) if there is a garage, put it at the side, near the front entrance.

9. Closing streets and alleys can help reduce crime because criminal outsiders are less likely to become familiar with the area, residents learn who does not belong in the neighborhood, residents committing crime in their own neighborhood cannot easily blame outsiders, burglars cannot easily gain access to properties, escape routes for robbers are blocked off, and drive-by shootings are prevented because cars cannot easily enter a street.

10. Traffic calming is used to change the environment of the street while still providing access to people. It includes installing road humps, narrowing the road, marking the road to create the illusion that it is narrowing, planting foliage along roadsides, building traffic circles and roundabouts, building traffic islands, installing gateways to residential neighborhoods, and permitting parking on both sides of residential streets.

11. Active place managers are important for crime prevention in addition to considerations of physical design.

12. Research has shown success in implementing design and management strategies at a wide variety of specific location types, including convenience stores, transit terminals, and universities/schools.

PROJECTS AND CHALLENGES

Interview project. Talk to an architect or planner about whether or not crime enters into decisions and designs.

Media project. Find an article about crime in a commercial or downtown area. Does it discuss the physical environment and its effect on the crime? Why or why not?

Map project. Map a park where some people seem to have driven out or scared off others. Who hangs out where and when? Examine how the park relates to the area surrounding it. How are the bushes and trees located and trimmed? What suggestions would you make for redesigning the park for wider and safer use?

Photo project. Take photos of four residential buildings to compare how some invite burglaries more than others. Be sure to take photos of the features you will focus on in your discussion.

Web project. Find a design for a future project (e.g., residential area, high school, city building, commercial complex, etc.). Do you think this project will generate more crime or less? Do not be taken in by the pretty picture.

NOTE

1. Many people view Jane Jacobs and Oscar Newman as offering conflicting ideas. Yet their *policies* were very close, even though they differed *rhetorically*. Both of them recognized that supervision of space is essential for a secure urban environment.

TEN

SITUATIONAL CRIME PREVENTION

G reat Britain's Home Office is roughly equivalent to the U.S. Department of Justice. Within this agency was a small research unit, located during the 1970s at Romney House on Marsham Street, a 5-minute walk from Scotland Yard. There, in 1973, a 31-year-old research officer named Ron Clarke had just completed a study of why youths abscond from borstals (American translation: why juvenile delinquents run away from reform school).

The usual social science variables did not successfully explain why some boys ran away whereas others stayed put. But Clarke learned that most boys ran away on weekends, when staffing and supervision were light. Because these were not prisons and staff members were not guards, their influence was largely informal. Merely by their presence, adults could prevent a certain amount of trouble, including absconding. With these results, Clarke began to think of crime in general as the result of human situations and opportunities.[1]

In 1976, with Pat Mayhew, A. Sturman, and J. M. Hough, Clarke published *Crime as Opportunity,* which explained many inexpensive ways to reduce crime by removing the opportunity to carry it out. Over time, this has become known as *situational crime prevention.* Clarke later headed the Research and Planning Unit of the Home Office. Under his leadership, several British researchers inside and outside the government created or discovered real-life crime prevention experiments that helped provide a major alternative theory of crime and practical guidelines for its prevention.

Clarke has encouraged or assisted others to study situational crime prevention examples with systematic data and to write up these studies. As it has evolved, situational crime prevention today includes at least 25 categories of prevention (Cornish & Clarke, 2003) and perhaps more than 200 case studies. Situational crime prevention seeks inexpensive means to reduce crime in three general ways:

1. Design safe *settings*. That includes the many methods presented in the previous chapter.

2. Organize effective *procedures*. That includes planning and carrying out the best management principles.

3. Develop secure *products*. That means making cars, stereos, and other products more difficult to steal or abuse (Clarke & Newman, 2005).

Indeed, the crime prevention repertoire is growing so greatly that it offers alternatives should one measure be politically or ethically problematic (see Felson & Clarke, 1997b; von Hirsch, Garland, & Wakefield, 2000). Settings, procedures, and products cover a wide range of crime prevention ideas, which no one person could learn in an entire lifetime. With Clarke's and others' multitude of examples, it is no longer possible to dismiss situational crime prevention as simply installing a better lock. Certainly, this field has produced many subtleties and surprises, dozens of books and monographs, and hundreds of articles (see the POP Center Web site, www.popcenter.org, for the Situational Crime Prevention Evaluation Database, a collection of articles about evaluations of situational crime prevention initiatives; for review of research literature, see also Clarke, 2004; Clarke & Eck, 2005; Clarke & Newman, 2005; Guerrette & Clarke, 2003; Knuttsson & Clarke, 2006; Maxfield & Clarke, 2004; Newman & Clarke, 2003; Smith & Clarke, 2000).

It is increasingly evident that situational crime prevention offers society the best chance for a quick and inexpensive way to reduce crime slice by slice. Thus, Clarke provides not only specific examples, but also principles for inventing your own crime prevention measures. Recently, Clarke and colleague Graeme Newman (2006) have applied the principles of situational crime prevention to terrorism in their book *Outsmarting the Terrorists.*

SITUATIONAL CRIME
PREVENTION AND CRIME ANALYSIS

Clarke and his associates adopted the following policy:

- Do not worry about academic theories. Just go out and gather facts about crime from nature herself (i.e., by observing, interviewing offenders, etc.). (This is not to say you should throw all your education to the wolves. It merely tells you that science has to gather facts and learn from them.)
- Focus on very specific slices of crime, particularly behaviors and environments (as discussed in Chapter 2), such as vandalism against telephones or soccer violence. Even the crime of "vandalism" would be far too broad!
- Do not try to improve human character. You are certain to fail.
- Try to block crime in a practical, natural, and simple way, at low social and economic cost.
- Do small-scale experiments, especially looking for natural environments (see Chapter 11) in which to study each slice of the crime prevention puzzle.
- Use very simple statistics and charts that let you see each comparison directly.

Perhaps we could sum up his approach in three words: "Don't get fancy."

Clarke sometimes claims that he really has no interest in theory, and that his only goal is to find practical ways to prevent crime. This surprises many conventional criminologists, but being practical poses a very good discipline on us all. Make it work! If it does not work, it probably is not very good science in the first place. If it does work, science will improve, too.

Another reason that situational crime prevention is a contribution to crime analysis is that it helps us understand offenders, targets, guardians, and their convergences. Clarke seeks to accomplish prevention by making each criminal act appear

- Difficult
- Risky
- Unrewarding

That breaks down crime into components that can then be explored—exactly what science is all about.

PREVENTING PROPERTY CRIME

A good deal of this chapter presents specific examples of successful situational crime prevention. They have been selected to tell a story. Included are crime prevention methods that were discovered accidentally, those involving criminologists, and others involving people who never heard of situational crime prevention, but did it anyway and were successful! Whether planned or not, people have acquired a variety of crime prevention experience well worth sharing.

Trouble on Double-Deck Buses

Our illustration of situational crime prevention begins with the problem of vandalism against Britain's traditional red double-deck buses. The Home Office researchers (Clarke, 1978) learned that most of the vandalism was on the upper deck, usually in the back row, where supervision was least likely to occur. They also learned that the traditional British bus conductor had a major role in preventing vandalism. A bus conductor would ascend the stairs to the upper deck to collect fares and thus serve as a guardian against the crime of vandalism.

Because some companies had removed the conductor to save money, whereas other companies had not, this was a natural experiment. Those buses with conductors had less vandalism, but they also had more assaults on conductors. This is an instance of how crime prevention can sometimes backfire, solving one crime but leading to another. This example also establishes that situational crime prevention is far from obvious, sometimes producing unexpected results.

Correcting the Criminal Use of Telephones

Ronald Clarke and associates have developed a growing literature on the criminal side of telephones and what to do about it (Exhibit 10.1). They have shown that obscene phone calls can be thwarted by caller identification services; drug transactions are impaired by pay phones that only call out; fraudulent international calls from pay phones are impossible when phones exclude common paths for the fraud; and stolen or cloned cell phones can be designed to fail for anybody but the owner. Clarke, Kemper, and Wyckoff (2001) documented more than $1.3 billion in cell phone fraud losses during 1995 to 1996. Six technical changes were designed to cut off fraud quickly:

1. Computer profiling to detect strange call patterns

2. Personal identification numbers (PINs)

3. Precall validation by computers

4. Operator checks

5. Radio wave checks

6. Encrypted checks of each phone

These adjustments resulted in a 97% cut in cell phone fraud.

Exhibit 10.1 Phone-Related Crime and Situational Solutions

Phone Crime Problem	Technical Solution	Reference
a. Obscene phone calls	Caller ID	Clarke, 1997a
b. Drug transactions	Only call out	Natarajan, Clarke, & Johnson, 1995
c. Fraudulent long-distance from pay phones	Programmed to exclude common frauds	Bichler & Clarke, 1996
d. Stolen or cloned cell phones	Designed to fail when stolen	Clarke, Kemper, & Wyckoff, 2001
e. Telephone vandalism	Remove or harden targets	Challinger, 1992

Telephones are important facilitators in drug transactions. Mangai Natarajan, Ronald Clarke, and Mathieu Belanger, in ongoing work, are paying close attention to the use of telephones for doing illegal work. Some localities have thwarted outdoor retail drug dealing by having pay phones

- Moved inside of businesses for extra supervision
- Programmed to call out but not receive calls
- Removed entirely

Car Theft Is Preventable

The interesting case of steering wheel locks preventing car theft already was offered in the Chapter 8 discussion on displacement. Additional information about thwarting motor vehicle theft is found in several studies (Brown, 1995; Brown & Billing, 1996; Southall & Ekblom, 1985). Clarke and Harris (1992) listed numerous technical changes that the auto industry can contribute to help reduce auto theft. Several of these are already common in cars today. Many cars have better security locks for steering columns, doors, and the hood. Door buttons today are more difficult to pull up with a clothes hanger. Window glass is often harder to break. Many models make it difficult to leave your keys in the ignition. Smart keys, elimination of external keyholes, and electronic immobilization after break-ins are no longer confined to the most expensive models.

Manufacturers have improved some of those models listed as most stolen by the Highway Loss Data Institute (see Exhibit 2.3). Tremendous strides in car stereo security have combined with lower fence values, thereby interfering with their theft. The time it takes to steal a car has increased, and the pure amateur has more problems than ever. Brown and Billing (1996) show that more secure cars lead to less theft in Britain, and the American auto industry experience shows that cars with disastrous theft problems can be redesigned for crime prevention and their good names restored. By the time you read this, a new design will have been developed, probably for a model that got into the national media as thieves' favorite.

On the other hand, cars have expensive gadgets or parts to steal, such as GPS units or catalytic converters, both of which have been hot items in recent years. Another example is airbags, which are quickly pried out and sold for about $1,000 for installation in cars at repair shops (for cars where the bags were deployed or even stolen!). This illustrates what Ekblom refers to as an "arms race" between offenders and forces of crime control. Crime is never permanently prevented, but neither do we get anywhere against crime when we do not try.

Beyond the automobile industry, inexpensive technology already exists to put a personal identification number into every new and valuable electronic item, such as a plasma television or DVD player. The product would not work outside your home unless you entered the right number. It would lose its value to a thief. It also should be possible to program something within your electrical system so an appliance removed from your home would not work elsewhere without punching in the code. Industry could make a major contribution to society by designing and selling more products that go kaput when stolen (see Clarke, 1999, 2004; Clarke & Newman, 2005; Felson, 1997).

A Serendipitous Finding About Motorcycle Theft

American motorcyclists keep complaining about having to put on their helmets and campaigning to stop helmet laws. If they only knew. Wherever helmet safety laws were enacted and enforced, thefts of motorcycles went down greatly.

To understand why, note that many motorcycle thefts are for joyriding and occur on the spur of the moment. The likely offender usually does not have a big motorcycle helmet with him at the time he sees a shiny motorcycle. When Germany enacted and enforced its motorcycle helmet law, thefts went down and stayed down, with no indication of displacement to other vehicle theft (Mayhew, Clarke, & Eliot, 1989).

We see that significant crime prevention can occur completely without planning. Even a very simple change in the law can have a great impact. Because wearing a helmet is highly visible behavior, it provides tangible evidence that the law is being followed and that the motorcycle probably is not stolen.

Saving Billions on Retail Theft

Not all prevention occurs with across-the-board laws enacted centrally. Some crime prevention requires more "personal service." For example, a retail store has to take into account its particular doors, layout, pedestrian flow, and hours of operation in planning for prevention. Good management and crime prevention go hand in hand within retail stores. A well-managed and well-organized retail store will not only have less shoplifting and employee theft but will usually enjoy more sales and better morale among employees (Clarke, 2002).

Retail stores use many prevention methods. More frequent inventories and audits help to discourage employee theft. Requiring that all merchandise be put in plastic bags that cinch at the top instead of in large open paper bags with large handles makes it harder for a customer to slip something unpaid for into his or her bag. Designing exit routes carefully encourages people to pay for their merchandise as they walk out. Tags that beep when not deactivated discourage shoplifters. To reduce fraudulent returning of items, major department stores put a separate sticker with a scanner code on every price tag at purchase. The sticker is scanned along with the price tag, so any clothes returned have to have that exact yellow sticker. Electronic systems for detecting merchandise are increasingly available at low prices, paying for themselves in loss reduction within a year or two. Robert DiLonardo's (1997) evaluation shows that tags can be tremendously successful in reducing thefts from stores. Barry Masuda (1993, 1997) shows that employee theft also can be reduced.

Retailers can easily lose thousands of dollars in merchandise out the door. In a few seconds, thieves can grab stacks of expensive garments and run to a waiting car. The well-managed store combines comprehensive planning with situational crime prevention to prevent such losses. For the back door, it is essential to schedule deliveries carefully so people do not take away more than they deliver. For the front door, a clever merchant learned to alternate the directions of hangers on the rack so they lock when grabbed. This small but ingenious idea is clearly superior to letting people steal and then waiting for the criminal justice system to find and punish them.

Our knowledge about retail crime has increased greatly in recent years (see Beck & Willis, 1995, 1999; Clarke, 2002; Gill, 1994; Hayes, 1997a, 1997b). A broader field of business crime analysis is offered in two collections of essays (Felson & Clarke, 1997a; Felson & Peiser, 1998). As you read these sources, you will realize that crime prevention should not simply be left to the public sector, although public officials can do an excellent job of preventing crime when they put their minds to it.

Refusing to Accept Subway Graffiti

For many years, the subway trains of New York City were covered inside and out with graffiti and surely were among the ugliest anywhere. Moreover, the transit system was in chaos, ridership was dropping, and employee morale was low. Many efforts and policies had failed to correct the problems.

Then David Gunn became president of the New York City Transit Authority and announced the Clean Car Program. The aim of the program was to clean off graffiti immediately. Graffiti painters thus would get no satisfaction from their work traveling all over town. New York City's subway cars never returned to the graffiti levels before the program (see Sloan-Howitt & Kelling, 1997). One lesson of the program: Find out exactly what potential offenders want from crime and take it away from them.

Another subway system far distant from New York City prevented graffiti in fixed locations using a very different plan. The Swedish government calls the Stockholm Metro the world's longest art gallery. More than half of its stations have artwork, including mosaics, paintings, engravings, and bas-reliefs. They may not win aesthetic fame, but the artists knew how to beat the graffiti painters with textures and colors. Each of these techniques was used: multicolors, surfaces that are either unusually rough or highly polished, and walls that were either sharply uneven or blocked with metal grills.

Art Theft Appreciation

Art theft is surprisingly common in New York City art galleries. Truc-Nhu Ho (1998) studied 229 such thefts from 45 art dealers. Although the statistics are limited, they show that art thefts fit patterns (see also Conklin, 1994, on routine activities and art theft; James, 2000). Art thieves

- Detest abstract art
- Avoid galleries with security checks
- Hate galleries near active nightlife
- Turn up their noses at large objets d'art
- Appreciate realistic paintings and sculptures
- Prefer galleries on the ground floor on quiet streets
- Resonate with art that has price tags affixed

The discerning art dealer should study art through the eyes of thieves.

Putting Lighting Into Focus

It is not so simple to say "Turn on the lights."

In the 1970s, it was very common for cities to fight crime by scattering streetlights without plan. Consider the logic for why this failed (Pease, 1999):

- Criminal activity is concentrated at or near specific places or blocks (Eck & Weisburd, 1995; Weisburd et al., 2004).
- Streetlight campaigns have often led to scattering placement without plan.

Totally *unplanned* lighting had little effect on crime. As a result, some analysts went to an extreme position, claiming that lighting cannot reduce crime. Ken Pease (1999) refers to these people as the "disciples of darkness."

Yet Painter and Farrington (1997) produced a rigorous study, with victim surveys showing a 41% reduction in crime in the lighting-enhanced area, compared to a 15% reduction in the control area. We have to conclude that lighting has a major possible contribution to reducing crime.

At the same time, lighting can *increase* crime in some cases. Lights can help a burglar see what he is doing. Lights can draw students back to school for after-hours vandalism. Lights can glare in the eyes of victims or guardians. Lights can make a better hangout for getting drunk and becoming disorderly. Thus, lights should not be placed without thought. Lighting can be highly effective in reducing crime when it is *clearly focused on the problem at hand* (Clarke, 2008; Painter & Farrington, 1997; Painter & Tilley, 1999; Pease, 1999).

In an excellent intellectual and factual review of the topic, Pease (1999) noted that a number of cities with strategic improvement of lighting clearly showed decreased crime rates. He also worked out how to think about lighting and to disaggregate the mechanisms whereby it might affect crime. Exhibit 10.2 shows his 17 different ways in which lighting can affect crime. The exhibit explains why lighting can lead to either more crime or less. It also shows that lighting can, surprisingly, affect crime in the daytime. For example, lights can give cues, even in daytime, that the area is not good for crime. It can keep people from moving out of the area, with fewer "for sale" signs to assist burglars in finding empty places to break into. After reading Exhibit 10.2, try to defend the position that the relationship between street lighting and crime is not a sophisticated enough topic for those of us in higher education to study.

Exhibit 10.2 Lighting Affects Crime in Many Ways

A. How More Lights Might *Reduce* Crime After Dark

1. Get people to spend late time in the yard or garden, serving as guardians.
2. Encourage people to walk more after dark, serving as guardians.
3. Make offenders more visible to guardians.
4. Make police on patrol more visible to offenders.

B. How More Lights Might *Increase* Crimes After Dark

1. Draw people away from home, assisting burglars.
2. Give offenders a better look at potential targets of crime.
3. Assist offenders in checking for potential guardians against crime.
4. Get nearby areas to *seem* darker, helping offenders to escape into them.

C. How More Lights Might *Reduce* Crimes in Daytime

1. Put new guardians on the street, those installing and maintaining lights.
2. Show official commitment; local citizens then cooperate in crime prevention.
3. Give cues—even in daytime—that the area is not good for crime.
4. Provide a talking point for citizens, who then get to know one another.
5. Keep people from moving out (fewer "for sale" signs to assist burglars).
6. Apprehend more offenders after dark, with fewer left for daytime offending.

D. How More Lights Might *Increase* Crimes in Daytime

1. Make it easy to pretend to be an electrical or maintenance employee.
2. Provide more nighttime fun that carries over to daytime drunkenness.
3. Set up new nighttime hangouts that might spill over as daytime trouble spots.

SOURCE: Adapted from Pease, K. (1999). A Review of Street Lighting Evaluations: Crime Reduction Efforts. In K. Painter and N. Tilley (Eds.), *Surveillance of Public Space: CCTV, Street Lighting and Crime Prevention*. Monsey, NY: Criminal Justice Press.

Music and Control

People are influenced not only by what they see but also by what they hear. Young people generally do not like classical music and will go away when it is played. That's far better than nightsticks and imprisonment. Music is also suitable for calming people down, as wise disc jockeys well know. When music is aggressive, crowds in bars are rowdiest (Scott & Dedel, 2006). The type of dancing also has a major influence on their behavior, with wilder dancing making people bump and, sometimes, fight. Yet the topic of music and crime has been little studied. Psychology students with expertise in perception and human factors are especially likely to break new ground in explaining how music provides cues that affect criminal behavior.

Situational Degeneration

Not only can crime situations be improved, but they can also be exacerbated. Thus, a store manager can remove crime control measures and cause shoplifting to rise. A homeowner can let well-trimmed bushes grow up, to the benefit of local burglars. A car manufacturer can cut costs by putting in cheaper steering wheel locks. One of the challenges of crime analysis is to put situational prevention and *situational degeneration* within the same intellectual framework. There is no better place to start than the study of violence.

PREVENTING VIOLENT CRIME

It is quite a mistake to think that situational crime prevention applies only to property crime. Understanding situational features of violence has grown considerably in recent years. The greatest source of progress stems from recognizing that violence is goal oriented and responds to cues from physical settings. As Chapter 3 explained, a book by James Tedeschi and Richard Felson (1994) shows us that all violence is goal oriented. A person might use violence (a) to get others to comply with wishes, (b) to restore justice as he perceives it, or (c) to assert and protect his self-image or identity. (As we shall see, these goals often make violence highly amenable to situational crime prevention as well.) A simple robbery starts out with the robber demanding your money and using or threatening force to get it. The robber is simply getting you to comply with his wishes—receiving your money without an argument. But if you challenge the robber in front of his co-offender, he may harm you to assert and protect his own identity (the third reason for violence). That is why it is best not to have a big mouth when someone is pointing a gun at you (see situational degeneration, above). It's also best not to go around giving people grievances against you; they may decide to restore justice. Fights between drunken young males usually occur as attempts to assert and protect identity. Road rage is often an effort to meet the second goal, restoring justice. Domestic violence can meet all three purposes (see R. Felson & Outlaw, 2007).

Even with predatory violence, although generally oriented toward the first purpose—gaining compliance—offenders will sometimes seek to protect identity or restore justice. For example, youths angry at the store owner who yelled at them may rob him not only for loot but also to retaliate and punish. Remember, all these evaluations are based on the *offender's viewpoint.* To understand violent or nonviolent crime, we cannot be distracted by our own moral outrage, or by the legal code, or by objective facts about what a person *ought* to think of others. If the guy in the bar hit you because he *thinks* you insulted him, the fact that he heard you wrong is entirely beside the point.

You might readily guess that alcohol plays a major role in violence. It gives people big mouths and big ears. Big mouths help people make aggressive statements that provoke counterattacks and restoration of justice. Big mouths also help people to provoke others into fights. Alcohol makes bigger ears by getting people to hear things that were not said. Managing alcohol is part of preventing violence (see Scott & Dedel, 2006).

Sports Events and Revelry

Speaking of alcohol, British football (soccer) has an unfortunate pattern of serious—and sometimes fatal—violence. Many fans arrive hours before a game, get drunk, and then commit acts of violence, many against fans of the visiting team. Because most of those involved in the violence do not own cars and therefore take buses to the games, the government arranged for these buses to arrive at the game later than in the past, allowing only a few minutes to buy a ticket and no time to get drunk. The effect was a reduction in football violence (Clarke, 1983).

Sweden also has a problem with alcohol-related violence, especially on one day each year. Midsummer's Eve (usually June 21) is the longest day of the year. In much of Sweden, this day has 24 hours of light. It is the most important holiday of the year. Swedes are usually reserved people, but they make an exception on Midsummer's Eve. A common behavior pattern is to get drunk and run wild. People also start bonfires, which sometimes get out of hand and burn more than intended. Moreover, many assaults occur on Midsummer's Eve. The crowds are far larger and wilder than anything police can handle, so deterrence loses its credibility. A more sensible policy was planned by Swedish authorities: They provided bonfires in designated and advertised locations and sought to channel the holiday spirit into these settings. Their efforts paid off by reducing assaults and other illegal behavior (see Bjor, Knutsson, & Kuhlhorn, 1992).

Compared with events like football games in Britain, American sports venues usually are not bad. The probable reason is that American teams try to sell a lot of tickets to families and business groups. This results in people of mixed ages and both sexes. Even in hockey, with its violence on the ice, there is reasonable peace in the stands. We all know of exceptions, but the rule remains.

American sports venues try to prevent people from bringing in their own bottles. This probably is so that they can sell more drinks, but they also use security justifications. They generally sell soft drinks and beer to the larger crowd, with hard drinks sold only within the corporate boxes. Beer sales are cut off later in the game, when some fans are a bit too drunk. Security people with binoculars keep an eye on the crowd to see if there are fights or if fans are getting dangerous. They then cut off the beer sales in that section or even start watering down the beer. Because beer is highly profitable to management, cutting off beer sales reduces proceeds, but it clearly enhances safety. Watering the beer gets the heavy drinkers to complain, but management is glad to give them their money back and have the drinking dwindle.

To prevent conflicts and fights when people are going out of a stadium, the strategy is to keep people moving, whether in cars or on foot, so they have little time to linger or to get mad. A well-managed stadium looks for bottle-necks where crowds cannot move, relieving the traffic problem quickly as a service to customers and as a way to prevent trouble (for a concise summary of literature and evaluation of strategies for spectator violence in stadiums, see Madsen & Eck, 2008).

Cruising

In many European and Hispanic nations, young people walk around the center of town on weekend evenings. The United States version of this activity is cruising in cars. Cruising creates traffic jams and interferes with business. The automobile spreads adolescent activity over more space and makes it harder to prevent trouble; thus, vandalism and assaults become more serious (see Felson, Berends, Richardson, & Veno, 1997; Wikstrom, 1995). Many U.S. cities have enacted special cruising ordinances or enforce traffic and parking ordinances more heavily in trying to control cruising (for a concise summary of literature and evaluation of strategies for cruising, see Glensor & Peak, 2004).

As explained by authors John Bell and Barbara Burke (1992), the city of Arlington, Texas, found that cruising by more than 1,000 cars was creating a major traffic jam on its main street for hours at a time. Ambulances could not get to hospitals, and little else in the way of normal city business could happen. Conventional traffic control methods were doing little good.

City Councilman Ken Groves learned that teenagers wanted two things: an unstructured and unsupervised environment in which to mingle, and restrooms. He speculated that if these were provided, most teenagers would act reasonably. A "cruising committee" was formed to link local agencies, businesses, the University of Texas at Arlington, and teenage representatives.

The committee devised a plan for the city to lease a large parking lot from the university and open it to cruisers on weekend nights while providing unobtrusive police protection, portable restrooms, and cleanup the next morning. Within two weekends, the new cruising area was in use by 1,000 parked or circling cars. The program channeled cruising into a smaller and safer area and pleased both teenagers and adults, while providing the gentle controls of a few police officers on the side.

The lesson of the program is that a crime problem may be related to another problem; solve the other problem, and the crime problem takes care of itself. In this case, the problem was to provide youths with an outlet for a social need in the context of the local situation. When this was done, the related crime problems dissipated.

Foul Play in College Water Polo

Many situational crime prevention measures emerge entirely by accident. An interesting example has to do not with a "crime" as such but with rule violations in the game of water polo. A former student was a water polo coach at the collegiate level and explained quite frankly how to cheat. When a member of the other team is about to get the ball or move toward the goal, simply put your hand inside his bathing suit, and he cannot proceed. This common form of foul play happens entirely under water, where the referees often fail to see it. The incentives to foul are strong and the controls are weak.

Water polo play got quite a bit cleaner some years ago. This did not happen because of more punishment or because players underwent moral regeneration; rather, new chemicals made the pool water less murky, so rule violations were easier to detect. As pools got clearer, water polo play got cleaner.

Barhopping and Bar Problems

On any given weekend night, more than 6,000 people from the surrounding towns and suburbs would go into Geelong, Australia, to socialize and drink alcohol. Some groups, drunk on the streets, would commit thefts or get into fights with one another. A typical pattern was this:

1. Drive to a packaged liquor outlet to purchase beer.

2. Drink beer in the car for an initial effect.

3. Go to the nearest bar for special prices.

4. Move to the next bar for its specials.

5. Go back to the car and drink more.

At this point, some people would use empty bottles as missiles to throw at people or property. The bars not only involved males in these efforts but also gave free drinks to young females to attract males. As the situation got worse, there were attacks on pub personnel. Bars worried about the money they lost by offering so many specials.

The police decided to do something and got the bar owners or managers together with the liquor board. They formulated "The Accord," a set of policies to discourage barhopping and other alcohol-related problems. It had more than a dozen provisions, but the most important were these:

- Cover charges to enter bars after 11:00 p.m.
- Denial of free reentry after someone exits

- No free drinks or promotions
- No extended happy hours
- A narrower drink price range
- Enforcement against open containers on the street

The Accord was a success in removing most of the street drinking and pub hopping, while reducing the violence and other crime problems in the central city (Felson, 1997).

Other important insights are provided by Ross Homel, Marg Hauritz, Gillian McIlwain, Richard Wortley, and Russell Carvolth (1997) in their study of drunkenness and violence around nightclubs in Surfer's Paradise, an Australian tourist resort. Tourists generate a lot of crime victimization and offending alike (see Pizam & Mansfield, 1996; Stangeland, 1995). The problems and policies that Homel's group discusses, however, can apply to any entertainment district. Among the alcohol policy features considered were

- Reduction of binge-drinking incentives, such as happy hours
- Low- and non-alcohol drinks and lower prices for them
- Staff policies to avoid admitting intoxicated persons
- Food and snacks available more of the time
- Varied clientele, not just hard drinkers
- Smaller glasses or drinks not as strong
- Strategies for dealing with problem customers
- Security training

The result was a substantial reduction in drunkenness and violence around the nightclubs.

Perhaps it is not surprising that a surgeon would be most aware of the ugly injuries from bar glasses. Jonathan Shepherd and his colleagues (see Shepherd, Brickley, Gallagher, & Walker, 1994) have written about the injuries reported by bar staff, classified by different types of glass. A straight-sided 1-pint glass produced 52 of 78 incidents. Only one of these injuries came from a splintered plastic glass. Tankards led to fewer injuries than straight-sided glasses. Half-pint glasses led to fewer injuries, but those drinking half a pint probably were not getting as drunk. Shepherd and colleagues (Shepherd, Hugget, & Kidner, 1993) also carried out an interesting experiment.

By collecting samples of different glass types and smashing them, they learned how nasty a weapon each produces. They found clearly that tankards are more difficult to smash and that tempered beer glasses break into a pile of relatively harmless chunks.

Making sure that bars use safer glasses is an example of what Clarke (1997b) calls "controlling crime facilitators." By paying close attention to what tools or weapons facilitate crime, we acquire more tools for preventing crime.

The general potential for regulating drinking environments to reduce crime has been discussed in an essay by Tim Stockwell (1997). In addition, Stuart Macintyre and Ross Homel (1997) offer a remarkable study titled "Danger on the Dance Floor." In examining behavior and accidents within discos and other nightclubs, they observed brushing, bumping, knocking, spilling drinks, pushing, shoving, hitting, and fighting. They found that the density of activities within nightclubs and the indoor design—including the location of tables and stools, pillars, walls, and bars, as well as the presence of disk jockeys—was very important. This is a good example of how situational crime prevention and crime prevention through environmental design intersect (for a concise summary of literature and evaluation of strategies for assaults in and around bars, see Scott & Dedel, 2006).

PREVENTING DRUNK DRIVING

Liquor policies influence not only intentional violence but also drunk driving and any accidental damage to property or people. H. Laurence Ross offers a brilliant analysis (1992) of how liquor policies and abuses are linked to drunk driving and subsequent deaths in his book *Confronting Drunk Driving*. Ross offers many surprising facts:

- Most drunk drivers involved in accidents or fatalities have never been arrested before for drunk driving. That means that "getting tough" on drunk drivers has its limits for preventing deaths.
- Upping the punishment levels has not accomplished anything in the past and probably will not accomplish anything in the future.
- Modern American society is organized so that it is natural to drive to the bar and back, and hence to drive with a blood alcohol level over the legal limit.

We can prevent drunk driving deaths and injuries only with more focused policies. These include making roads and cars safer to prevent accidents or reduce the injury from them, or to use the regulatory system to get bars to stop serving people who are already drunk.

Australian and Scandinavian efforts to reduce drunk driving have been quite successful in many cases. These include random breath tests on highways (see Graham & Homel, 2008). In New South Wales, they have learned to give dramatic publicity to their breath testing, not only with media coverage but also by placing at the side of the road a large testing vehicle with a big sign reading "Booze Bus." Even the license plates have these words, helping to get people talking and reminding one another not to mix drinking and driving. The public responds quite well to these efforts and tends to reduce its drunk driving, without many arrests and with no draconian punishment.

American efforts to raise drinking ages and make them consistent among states also have produced a major decline in drunk driving and related injuries and deaths. American society has long had in place rules or laws against drinking in the streets and serving alcohol to those already drunk, and limiting the size and conditions of bars. Of course, they are not always enforced (for a concise summary of literature and evaluation of strategies for drunk driving, see Scott, Emerson, Antonacci, & Plant, 2006).

PREVENTING FRAUD

We are increasingly recognizing that situational crime prevention can help reduce fraud. Here are some important illustrations:

Bad checks. Knutsson and Kuhlhorn (1997) found that easy check cashing makes for easy check fraud. When rules were tightened, that crime declined significantly (just as Tremblay, 1986, found in Canada). When banks refused to guarantee bad checks, the merchants stood to lose money and started to be careful before they would hand out cash.

Misleading information. Kuhlhorn (1997) studied how people cheat the government by filling in conflicting information on different forms. Computer comparisons were made to reveal fraud, and the public was told about this development. As a result, people cheated much less often.

Illicit refunds. Many people defraud retail stores by stealing goods, convincing the store they were bought there, then getting a cash refund. Challinger (1997) showed that new rules for refunds made this type of fraud more difficult to accomplish.

Employee falsification. Most organizations that reimburse employees require original receipts to discourage fraudulent medical claims or expense reimbursements.

Embezzling employees. Well-designed auditing and accounting systems make it harder for one person to steal money from an organization. For example, when more than one person signs each large check and when independent auditors go over the books, less fraud occurs. Some people still conspire to commit fraud, but the whole idea of designing out fraud is to *require* conspirators for crime to be committed and hope one of them will lose his or her nerve.

Construction corruption. Racketeering in the New York City construction industry combines fraud with extortion, bribery, theft, sabotage, and bid rigging. The Organized Crime Task Force, directed by Ron Goldstock, involved James Jacobs of New York University and several others to analyze organized crime's involvement in construction. Their recommendations were to change the structure and industry characteristics generating the motivation, ability, and opportunity to act corruptly. They invented the ugly term "racketeering susceptibility," but more important, they realized that the very structure of the industry was creating racketeering opportunities. By altering that structure, organized crime could be made less likely to succeed (Organized Crime Task Force, 1988).

PREVENTING INTERNET FRAUD

Internet fraud includes any type of fraud scheme that uses one or more components of the Internet to present fraudulent solicitations to prospective victims, to conduct fraudulent transactions, or to transmit the proceeds of fraud to financial institutions or to others connected with the scheme. The components of the Internet might include email, chat rooms, message boards, and/or Web sites.

Identity fraudsters may seek to trick the victim into direct transfers of money or goods for a promise not delivered. Other fraudsters seek to steal information from the victim to be used to steal money indirectly via a third party, such as a credit card company. Still other fraudsters get the second party to provide information to assist in getting at a third party. Thus, a bank may pay the price for a fake account set up.

Internet fraudsters offer a variety of goods or services to those they reach, including free offers, participation in auctions, investments, business opportunities, "work-at-home" schemes, advanced fee loans, and more. Some deliver goods that happen to be counterfeit or fake pharmaceuticals or inoperable electronics.

As we speak, computer software companies are developing and improving software to prevent these crimes. Internet service providers increasingly screen out scams before they arrive. Software is now designed to help keep passwords private and to warn people as they open fraudulent emails or Web pages. Some scams are exposed as quickly as possible on the Internet.

Some credit card issuers offer "substitute" or "single-use" credit card numbers—these allow you to use your credit card without putting your real account number online. Many transactions require copying the security code on the credit card in addition to the other numbers.

Laws have been established that require businesses and institutions to protect private information better—examples include HIPAA (Health Insurance Portability and Accountability Act of 1996), and FACTA (Fair and Accurate Credit Transactions Act of 2003) (Newman, 2004). As part of FACTA, the Federal Trade Commission implemented the "Red Flags" rule in January 2008 which requires many businesses and organizations to implement a written identity theft prevention program designed to detect the warning signs of identity theft in their day-to-day operations, take steps to prevent the crime, and control the damage inflicted (Federal Trade Commission, 2009). Media and other crime prevention education sources inform people about how to protect their personal information, such as using credit cards (not debit cards) for transactions where the card leaves their sight. Even entire companies have been created whose main function is to constantly monitor people's credit scores to help protect individuals from serious identity theft (see Newman, 2003, 2004; Newman & Clarke, 2003).

PREVENTING REPEAT VICTIMIZATION

Queen Elizabeth bestowed the Order of the British Empire (O.B.E.) on criminologist Ken Pease for his contributions to crime prevention. Pease (1992; see also Farrell, 1995) had demonstrated that a very large share of crime victim-izations were "repeats." People victimized once are especially likely to be victimized again.

Pease figured out how to focus prevention on those already victimized. When someone's home was burglarized a first time, a prevention team would zero in on that particular unit to prevent a repetition. The team enlisted the res-idents of the five or six homes nearest the burglarized unit to keep an eye on it, a "cocoon" neighborhood watch. The unit also helped improve locks and doors, and otherwise reduce the risk. Those housing units in the experimental group saw declining risk of burglary. The unit's success was far greater than for the usual methods, such as the unfocused and ineffective neighborhood watch. Pease's focus on reducing repeat victimization is increasingly applied to other offenses (Anderson & Pease, 1997; Bowers & Johnson, 2004, 2005; Farrell, 1995; Farrell, Tseloni, & Pease, 2005; Johnson & Bowers, 2004a, 2004b, 2007; Townsley, Homel, & Chaseling, 2003). Its advantages include

- Efficiently reducing crime at low cost
- Avoiding the usual political controversies
- Assisting the worst victims
- Helping everyone think more clearly about crime

Students of crime should take note of major American efforts by the National Institute of Justice to prevent repeat victimization on this side of the Atlantic. By the time this book is out, results of these studies might be available.

PREVENTING THE SALE OF STOLEN GOODS

As explained in Chapter 5, markets for stolen goods are extremely important. Mike Sutton (1998) elaborated the "market reduction" approach to prevent theft and burglary. Detectives have long known to watch pawnshops, jewelry stores, auto body shops, even flea markets. Crime prevention specialists are beginning to devise more elaborate efforts at market reduction. A careful department of motor vehicles can interfere with registration of stolen cars, or with converting registrations of crashed cars to stolen cars of the same model. Requiring identification when getting cash for recycled metals such as copper can reduce metal theft. The Internet offers a fast way to circulate pictures of stolen jewelry to merchants. Repair contracts for electronics goods could readily be used to trace their ownership and thus help defeat theft. Computers can handle a lot of this effort, but the reality is lagging behind the potential.

CONCLUSION

Situational crime prevention offers a broad repertoire for preventing crime here and now, rather than there and eventually. It is verifiable, clear, simple, and cheap. It is available to people of all income groups, seldom treading on civil liberties (see Felson & Clarke, 1997b).[1] Situational crime prevention bypasses the hardliners and softheads. Its idealism is not utopian because it has found practical ways to do the right thing. Most often it applies to a narrow slice of crime, but sometimes it can be mass-produced effectively. Exhibit 10.3 shows how the process of control proceeds in six steps. First, we try to build human character. Then we design secure environments, as Chapter 9 explained. Next, we use other means to remove crime situations, as this chapter considered. Then we make arrests and process suspects, try and convict offenders, and punish and rehabilitate. We have made it quite clear that our

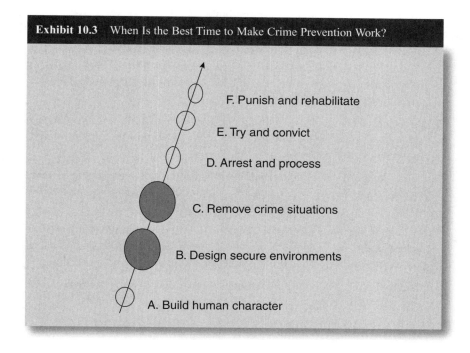

Exhibit 10.3   When Is the Best Time to Make Crime Prevention Work?

F. Punish and rehabilitate

E. Try and convict

D. Arrest and process

C. Remove crime situations

B. Design secure environments

A. Build human character

most realistic chance for reducing crime occurs during Steps B and C—designing secure environments and removing crime situations. In other words, situational crime prevention (broadly speaking) offers us our best chance to minimize crime, without interfering substantially or negatively with people's lives. As the repertoire of prevention methods continues to grow, we have a means for slicing away at crime.

MAIN POINTS

1. Situational crime prevention is highly focused on preventing crime here and now and on very specific slices (situations) of crime. It is practical, not utopian. Situational crime prevention seeks inexpensive means to reduce crime in three general ways: design safe settings; organize effective procedures; and develop secure products.

2. Situational crime prevention reduces the inducements to commit crime by making crime targets less rewarding while increasing the risk and effort associated with crime.

3. Situational crime prevention generally does not displace crime elsewhere. Indeed, crime prevention often leads to a "diffusion of benefits," reducing crime even beyond the immediate setting.

4. Specific examples of successful situational crime prevention for property crime include addressing vandalism on double-deck buses, correcting criminal use of telephones, preventing car and motorcycle theft, reducing retail theft, refusing to accept subway graffiti, as well as preventing fraud and sale of stolen goods.

5. Strategies such as utilizing lighting and controlling music are also effective examples of situational crime prevention, but these must be implemented in thoughtful, constructive ways.

6. Crime situations can be improved, but they can also be made worse by making changes to the settings, procedures, and products that increase opportunities for crime.

7. Specific examples of successful situational crime prevention for violent crime include reducing violence at spectator sports events, addressing cruising, controlling bar hopping and bar problems, and preventing drunk driving.

8. Identity theft is a new and increasing problem that is being addressed on a large and small scale with situational crime prevention techniques.

9. Advantages of addressing repeat victims include efficiently reducing crime at low cost; avoiding the usual political controversies; assisting the worst victims; and helping everyone think more clearly about crime.

PROJECTS AND CHALLENGES

Interview projects. (a) Talk to a security person in the retail field. Ask specific questions about each type of situational crime prevention. What does he or she prefer, use, or ignore? (b) During off-duty or slack hours, interview a bartender about specific methods used to prevent conflict from developing and escalating. Ask about shutting off those drinking too much, how to refuse those who are underage, and how to calm people down. What does he or she do when someone spills a drink?

Media project. (a) Check out the magazines in the security field. What products are advertised there, and what situational crime prevention methods are left out? (b) Find out whether any car manufacturer has made major efforts to reduce a certain model's vulnerability to theft. Then use the Highway Loss Data Institute pamphlets to see whether its theft rates really declined relative to other models.

Map project. Map out a shopping mall or mini-mall. Where are its weak spots and strong spots from a situational crime prevention viewpoint?

Photo project. Devise a low-cost situational crime prevention method to make a college dormitory more secure from crime. Cover as many types of situational crime prevention as you can, using photos to strengthen your argument.

Web project. Go to www.popcenter.org and read one of the problem guides. Using the response table at the end of a guide, think about which responses are shown to be most effective and which are based on situational crime prevention. Do they overlap?

NOTES

1. For more on the history of these efforts, see Sullivan (2000), von Hirsch et al. (2000), and Clarke and Felson (in press).

2. Some people make moral and political attacks on situational crime prevention, but any techniques raising ethical controversies are greatly outnumbered by the ones that are innocuous but effective.

ELEVEN

EVERYDAY TECHNOLOGY AND EVERYDAY CRIME

———⊷•⊶•⊷———

This book has offered a simple theory for understanding crime and how it changes. This theory is based on elements of everyday life:

A. Everyday crime,

B. Everyday routines,

C. Everyday technology, and

D. Small inventions.

The everyday crime is generated by everyday routines, which result from everyday technology. That technology changes in response to small inventions that occur from time to time and are implemented in real life. Not all inventions have an important impact—only those inventions that alter the daily routines and affect who does what, when, where, and how, alter crime involvement. This simple theory of crime change is depicted in Exhibit 11.1.

In developing these principles, we have drawn from diverse scholars, including urban planners, anthropologists, geographers, political scientists, economists, psychologists, sociologists, and business specialists. Despite various points of departure, all roads lead to a single place: Crime is a tangible

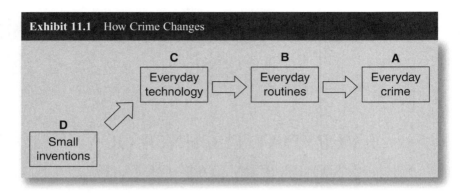

Exhibit 11.1 How Crime Changes

activity depending on other activities in everyday life. Clearly, the study of crime fits within a larger system of knowledge. The experience of practitioners in public and private sectors leads to exactly the same spot.

Throughout this book, we have shown how simple inventions affect crime opportunities and thus drive crime rates. This is entirely consistent with William F. Ogburn's (1964) theory of social change: Inventions are the major force driving human history. Technology should not be mistaken for "high technology" because human life has been changed greatly by simple innovations. Barbed wire is hardly highly complex, but it made possible the settling of the western United States by separating grazing grounds. Spraying the swamps wiped out many diseases borne by mosquitoes and helped triple the world's population. Washing hands before delivering babies completely transformed the infant mortality rate.

Thus, we can understand why something as simple as steering wheel locks transformed auto theft rates in Germany. In the 1960s and 1970s, lightweight plastics altered everyday products, making them much easier to steal. Crime rates during the 1990s reversed direction, largely responding to more widespread money machines and quicker point-of-sale transactions using plastic cards, removing much of the cash available to steal. The very recent acceleration of telecommuting could well place more guardians in residential areas and reduce their daytime crime. Monetary inventions provide a chance to steal, counterfeit, embezzle, abuse a credit card, or jump someone at the money machine. Newer innovations impair counterfeiting. Inventions affect how readily cars or electronic equipment can be stolen and point toward the future projection of crime problems (see Clarke & Newman, 2005; Davis & Pease, 2000; Felson, 1998).

Yet such projection should always bear in mind that the impact of new technology on crime always occurs through simple human processes:

- Slipping into an office to abscond with a password
- Stealing Web-ordered packages at the doorstep
- Hiding designer drugs as small pills inside a pocket
- Changing a number on the computer screen at work
- Waving a weapon toward your nose

Although very significant for changes in society—and its crime rates— few inventions stand alone as earthshaking contributions by a single inventor (Gilfillan, 1935). Incremental inventiveness is the key to how technology changes over time. Not only do people add a little here, a little there, to make something more useful over time, but the history of invention has more lags than leaps. For example, more than a half century has passed since John Bardeen and his colleagues at Bell Labs in New Jersey invented the transistor. It took at least 15 years for transistorized electronic goods to become commonplace (hence, suitable for widespread theft). It took more than 40 years of incremental change before cell phones and the Internet created new niches for illegal action. More trial and error was needed to devise antidotes for these crime problems. Inventions are the mother of change, but the progeny take time to grow up.

Do not think of inventions as gadgets alone. Inventions include innovations in business types, job responsibilities, software, ways to train people, credit card rules, methods for managing barrooms, money machine protocols, or environmental designs (see Chapters 9 and 10). Each of these could bring along new crime opportunities or render an old crime obsolete, such as better safes that defeat the efforts of safecrackers.

CONCLUSION

This book has sought to demonstrate that crime analysis is possible. We can gain clarity if we accentuate the tangible. That means sticking close to criminal events and specifying exactly how, when, and where offenders act; their specific motives; how noncriminal realities give rise to criminal events; and the step-by-step sequences that crime follows in daily life. In so doing, we can formulate a means to study the specifics of crimes and a very general theory of crime covering many different types and spanning nations and eras. In general, offenders find out about suitable crime targets via personal ties, shared activity spaces, and specialized roles. The last of these methods for finding a crime target is essential, not only for understanding the basis for "white-collar" crime but also for linking it to a larger theory for all other types of crime.

Crime is a process, depending on the convergence of offenders and targets in the absence of guardians. The transportation system generates these convergences. Markets for stolen goods are of central importance for generating crime opportunities, and opportunities make the thief. One crime affects the opportunity for another, teaching us to study "multiplier effects" in detail. Each specific type of crime carves its niche in the settings, procedures, and products of a larger society. By altering that niche, society can greatly reduce crime. Such efforts accomplish much more than inviting people to commit crimes and then trying to catch them.

In general, crime can be viewed as a routine activity that feeds on routine legal activities as people live their lives and proceed to earn their daily bread. Technology alters crime niches by changing crime targets themselves, how easily one can get to them, or how readily they can be attacked or protected. None of this can be understood without putting yourself into the mind of a criminal and the simplest of tasks he must carry out. As you do so, you will begin to understand. This talk of technology should not put you off, for its impact depends on the human touch. Most computer fraud is traced not to brilliant hackers but rather to those who leave their passwords on top of the desk with their office doors open, or who dump the technical manuals in the trash bin, or who used to work somewhere and still access information they shouldn't. People don't seem to change, even as machines, organizations, and locations are completely transformed (see Ekblom, 2002, 2005a, 2005b; Pease, Rogerson, & Ellingworth, 2008).

Crime is a set of tasks, and society sometimes makes those tasks easy and other times makes them hard. Therein lies the secret of how crime varies, how it changes, and how it can be prevented. Inventions are what determine daily tasks and hence are the driving force in the history of crime. Students of crime analysis should stop wasting their efforts looking deep into social structure or the human soul.

REFERENCES

———•◆•———

Altizio, A., & York, D. (2007). Robbery of convenience stores (Guide No. 49). Retrieved September 25, 2009, from http://www.popcenter.org/problems/robbery_convenience/

Anderson, D., & Pease, K. (1997). Biting back: Preventing repeat burglary and car crime in Huddersfield. In R. V. Clarke (Ed.), *Situational crime prevention: Successful case studies* (2nd ed., pp. 200–208). New York: Harrow & Heston.

Apel, R., Bushway, S., & Brame, R. (2007). Unpacking the relationship between adolescent employment and antisocial behavior: A matched samples comparison. *Criminology, 45*(1), 67–97.

Asbridge, M., Smart, R. G., & Mann, R. E. (2006). Can we prevent road rage? *Trauma, Violence & Abuse, 7*(2), 109–121.

Baldry, A. C., & Farrington, D. P. (2007). Effectiveness of programs to prevent school bullying. *Victims & Offenders, 2*(2), 183–204.

Bamfield, J. (1998). A breach of trust: Employee collusion and theft from major retailers. In M. Gill (Ed.), *Crime at work: Increasing the risk for offenders.* Leicester, UK: Perpetuity Press.

Bandura, A. (1985). The psychology of chance encounters and life patterns. *American Psychologist, 37,* 747–755.

Barker, R. (1963). *The stream of behavior.* New York: Appleton-Century-Crofts.

Barker, R. G., & Gump, P. V. (Eds.). (1964). *Big school, small school.* Stanford, CA: Stanford University Press.

Baron, S. W. (1997). Risky lifestyles and the link between offending and victimization. *Studies on Crime and Crime Prevention, 6*(1), 53–72.

Baron, S. W., Forde, D. R., & Kay, F. M. (2007). Self-control, risky lifestyles, and situation: The role of opportunity and context in the general theory. *Journal of Criminal Justice, 35*(2), 119–136.

Barron, O. (2000). Why workplace bullying and violence are different: Protecting employees from both. *Security Journal, 13*(2), 63–72.

Baumeister, R. F., Heatherton, T. F., & Tice, D. M. (1994). *Losing control: How and why people fail at self-regulation.* San Diego, CA: Academic Press.

Baumgartner, M. P. (1993). Violent networks: The origins and management of domestic conflict. In R. B. Felson & J. T. Tedeschi (Eds.), *Aggression and violence:*

Social interactionist perspectives (pp. 209–231). Washington, DC: American Psychological Association.

Beauregard, E., Proulx, J., & Rossmo, K. (2007). Script analysis of the hunting process of serial sex offenders. *Criminal Justice and Behavior, 34*(8), 1069–1084.

Beck, A., & Willis, A. (1995). *Crime and security: Managing the risk to safe shopping.* Leicester, UK: Perpetuity Press.

Beck, A., & Willis, A. (1999). Context-specific measures of CCTV effectiveness in the retail sector. In K. Painter & N. Tilley (Eds.), *Surveillance of public space: CCTV, street lighting and crime prevention* (pp. 251–269). Monsey, NY: Criminal Justice Press.

Beckett, K., & Sassoon, T. (2000). *The politics of injustice: Crime and punishment in America.* Thousand Oaks, CA: Pine Forge.

Bell, D. (1960). *The end of ideology: On the exhaustion of political ideas in the fifties.* Glencoe, IL: Free Press.

Bell, J., & Burke, B. (1992). Cruising Cooper Street. In R. V. Clarke (Ed.), *Situational crime prevention: Successful case studies* (pp. 108–112). New York: Harrow & Heston.

Bellamy, L. (1996). Situational crime prevention and convenience store robbery. *Security Journal, 7*(1), 41–52.

Benson, M. L., & Simpson, S. (2009). *White collar crime: An opportunity perspective.* New York: Taylor & Francis.

Bentham, J. (1907). *An introduction to the principles of morals and legislation.* Oxford, UK: Clarendon. (Original work published 1789)

Best, J. (1999). *Random violence: How we talk about new crimes and new victims.* Berkeley: University of California Press.

Bichler, G., & Clarke, R. V. (1996). Eliminating pay phone toll fraud at the Port Authority Bus Terminal in Manhattan. In R. V. Clarke (Ed.), *Preventing mass transit crime* (pp. 93–115). Monsey, NY: Criminal Justice Press.

Biron, L., & Ladouceur, C. (1991). The boy next door: Local teen-age burglars in Montreal. *Security Journal, 2*(4), 200–204.

Bjor, J., Knutsson, J., & Kuhlhorn, E. (1992). The celebration of Midsummer Eve in Sweden—A study in the art of preventing collective disorder. *Security Journal, 3*(3), 169–174.

Block, C. R., & Skogan, W. G. (2001). *Do collective efficacy and community capacity make a difference behind "closed doors"?* Chicago: Illinois Criminal Justice Information Authority.

Block, R. L., & Block, C. R. (1992). Homicide syndromes and vulnerability: Violence in Chicago's community areas over 25 years. *Studies on Crime and Crime Prevention, 1*(1), 61–85.

Block, R. L., & Block, C. R. (2000). The Bronx and Chicago: Street robbery in the environs of rapid transit stations. In V. Goldsmith, P. G. McGuire, & J. H. Mollenkopf (Eds.), *Analyzing crime patterns: Frontiers of practice* (pp. 137–152). Thousand Oaks, CA: Sage.

Blumberg, P. (1989). *The predatory society: Deception in the American marketplace.* New York: Oxford University Press.

Boba, R. (2008). *Crime analysis with crime mapping.* Thousand Oaks, CA: Sage.

Boba, R., & Santos, R. (2008). A review of the research, practice, and evaluation of construction site theft occurrence and prevention: Directions for future research. *Security Journal, 21*(4), 246–263.

Bouffard, J., Exum, M. L., & Paternoster, R. (2000). Whither the beast? The role of emotions in a rational choice theory of crime. In S. S. Simpson (Ed.), *Of crime and criminality: The use of theory in everyday life* (pp. 159–178). Thousand Oaks, CA: Pine Forge.

Bowers, K. J., & Johnson, S. D. (2003). Measuring the geographical displacement and diffusion of benefit effects of crime prevention activity. *Journal of Quantitative Criminology, 19*(3), 275–301.

Bowers, K. J., & Johnson, S. D. (2004). Who commits near repeats? A test of the boost explanation. *Western Criminology Review, 5,* 12–24.

Bowers, K. J., & Johnson, S. D. (2005). Domestic burglary repeats and space-time clusters: The dimension of risk. *European Journal of Criminology, 20,* 67–92.

Braga, A. A. (2003). *The problem of gun violence among serious young offenders.* Washington, DC: Office of Community Oriented Policing Services.

Braga, A. A. (2005). Hot spots policing and crime prevention: A systematic review of randomized controlled trials. *Journal of Experimental Criminology, 1*(3), 317–342.

Brame, R., Bushway, S. D., Paternoster, R., & Apel, R. (2004). Assessing the effect of adolescent employment on involvement in criminal activity. *Journal of Contemporary Criminal Justice, 20*(3), 236–256.

Brantingham, P. J., & Brantingham, P. L. (1975). The spatial patterning of burglary. *Howard Journal of Penology and Crime Prevention, 14,* 11–24.

Brantingham, P. J., & Brantingham, P. L. (1984). *Patterns in crime.* New York: Macmillan.

Brantingham, P. J., & Brantingham, P. L. (1994). Surveying campus crime: What can be done to reduce crime and fear? *Security Journal, 5*(8), 160–171.

Brantingham, P. J., & Brantingham, P. L. (1998). Environmental criminology: From theory to urban planning practice. *Studies on Crime and Crime Prevention, 7*(1), 31–60.

Brantingham, P. J., & Brantingham, P. L. (2003). Anticipating the displacement of crime using the principles of environmental criminology. In M. Smith & D. Cornish (Eds.), *Theory for practice in situational crime prevention* (pp. 119–148). Monsey, NY: Criminal Justice Press.

Brantingham, P. L., & Brantingham, P. J. (1984). Mobility, notoriety and crime: A study in crime patterns of urban nodal points. *Journal of Environmental Systems, 11*(1), 89–99.

Brantingham, P. L., & Brantingham, P. J. (1993). Nodes, paths and edges: Considerations on the complexity of crime and the physical environment. *Journal of Environmental Psychology, 13*(1), 3–28.

Brantingham, P. L., & Brantingham, P. J. (1999). A theoretical model of crime hot spot generation. *Studies on Crime and Crime Prevention, 8*(1), 7–26.

Brantingham, P. L., Brantingham, P. J., & Wong, P. S. (1991). How public transit feeds private crime: Notes on the Vancouver "Skytrain" experience. *Security Journal, 2*(2), 91–95.

Bratton, W., & Knobler, P. (1998). *Turnaround: How America's top cop reversed the crime epidemic.* New York: Random House.

Britt, C. L., & Gottfredson, M. R. (Eds.). (2003). *Control theories of crime and de linquency.* New Brunswick, NJ: Transaction.

Britt, T. W., & Garrity, M. J. (2006). Attributions and personality as predictors of the road rage response. *British Journal of Social Psychology, 45*(1), 127–147.

Brown, R. (1995). *The nature and extent of heavy goods vehicle theft* (Paper No. 66, Crime Detection and Prevention Series). London: British Home Office Research Publications.

Brown, R., & Billing, N. (1996). *Tackling car crime: An evaluation of sold secure* (Paper No. 71, Crime Detection and Prevention Series). London: British Home Office Research Publications.

Bureau of Justice Statistics. (2006). *Crime victimization in the United States, 2005.* Retrieved March 9, 2009, from http://www.ojp.usdoj.gov/bjs/

Bureau of Justice Statistics. (2009). *Crime and victim statistics.* Retrieved March 9, 2009, from http://www.ojp.usdoj.gov/bjs/cvict.htm

Burgess, E. W. (1916). Juvenile delinquency in a small city. *Journal of the American Institute of Criminal Law and Criminology, 6,* 724–728.

Burke, R. H. (1998). *Zero tolerance policing.* Leicester, UK: Perpetuity Press.

Burrows, J., Shapland, J., Wiles, P., & Leitner, M. (1993). *Arson in schools.* London: Arson Prevention Bureau.

Center for Problem-Oriented Policing. (2009). Retrieved May 26, 2009, from http://www.popcenter.org

Chaiken, J., Lawless, M., & Stevenson, K. (1974). *The impact of police activity on crime: Robberies and the New York City subway system* (Report No. 20R-1424-NYC). Santa Monica, CA: RAND.

Chainey, S. P., & Ratcliffe, J. H. (2005). *GIS and crime mapping.* London: Wiley.

Challinger, D. (1992). Less telephone vandalism: How did it happen? In R. V. Clarke (Ed.), *Situational crime prevention: Successful case studies* (pp. 75–88). New York: Harrow & Heston.

Challinger, D. (1997). Refund fraud in retail stores. In R. V. Clarke (Ed.), *Situational crime prevention: Successful case studies* (2nd ed., pp. 250–262). New York: Harrow & Heston.

Chambliss, W. J. (1999). *Power, politics and crime.* Boulder, CO: Westview.

Chin, K. (1999). *Smuggled Chinese: Clandestine immigration to the United States.* Philadelphia: Temple University Press.

Chin, K. (2001). The social organization of Chinese human smuggling. In D. Kyle & R. Koslowski (Eds.), *Global human smuggling: Comparative perspectives* (pp. 216–234). Baltimore: Johns Hopkins University Press.

Clarke, R. V. (1978). *Tackling vandalism.* London: British Home Office Research Publications.

Clarke, R. V. (1983). Situational crime prevention: Its theoretical basis and practical scope. In M. Tonry & N. Morris (Eds.), *Crime and justice: An annual review of research* (Vol. 4, pp. 225–256). Chicago: University of Chicago Press.

Clarke, R. V. (1997a). Deterring obscene phone callers. In R. V. Clarke (Ed.), *Situational crime prevention: Successful case studies* (2nd ed., pp. 90–97). New York: Harrow & Heston.

Clarke, R. V. (Ed.). (1997b). *Situational crime prevention: Successful case studies* (2nd ed.). New York: Harrow & Heston.

Clarke, R. V. (1999). *Hot products: Understanding, anticipating and reducing demand for stolen goods* (Paper No. 112, Police Research Series). London: British Home Office Research Publications.

Clarke, R. V. (2002). *Shoplifting.* Washington, DC: U.S. Department of Justice, Office of Community Oriented Policing Services.

Clarke, R. V. (2004). Crime profiling of products: The idea and the substance. *IEEE Technology and Society Magazine, 23*(3), 21–27.

Clarke, R. V. (2005). *Closing streets and alleys to reduce crime: Should you go down this road?* Washington, DC: U.S. Department of Justice, Office of Community Oriented Policing Services.

Clarke, R. V. (2008). *Improving street lighting to reduce crime in residential areas.* Washington, DC: U.S. Department of Justice, Office of Community Oriented Policing Services.

Clarke, R. V., & Eck, J. E. (Eds.). (2005). *Crime analysis for problem solvers in 60 small steps.* Washington, DC: U.S. Department of Justice, Office of Community Oriented Policing Services.

Clarke, R. V., & Eck, J. E. (2007). *Understanding risky facilities.* Washington, DC: U.S. Department of Justice, Office of Community Oriented Policing Services.

Clarke, R. V., & Felson, M. F. (in press). The origins of situational crime prevention and the routine activity approach. In F. T. Cullen & P. Wilcox (Eds.), *Encyclopedia of criminological theory.* Thousand Oaks, CA: Sage.

Clarke, R. V., & Goldstein, H. (2002). Reducing auto theft at construction sites: Lessons from a problem-oriented project. In N. Tilley (Ed.), *Analysis for crime prevention* (pp. 89–130). Monsey, NY: Criminal Justice Press.

Clarke, R. V., & Harris, P. M. (1992). Auto theft and its prevention. In M. Tonry (Ed.), *Crime and justice: A review of research* (Vol. 16, pp. 1–54). Chicago: University of Chicago Press.

Clarke, R. V., Kemper, R., & Wyckoff, L. (2001). Controlling cell phone fraud in the US—Lessons for the UK. *Security Journal, 14*(1), 7–22.

Clarke, R. V., & Mayhew, P. (1998). Preventing crime in parking lots: What we know and what we need to know. In M. Felson & R. B. Peiser (Eds.), *Reducing crime through real estate development and management* (pp. 125–136). Washington, DC: Urban Land Institute.

Clarke, R. V., & Newman, G. (Eds.). (2005). *Designing out crime from products and systems.* Monsey, NY: Criminal Justice Press.

Clarke, R. V., & Newman, G. (2006). *Outsmarting the terrorists.* Portsmouth, NH: Greenwood.

Clarke, R. V., & Smith, M. (2000). Crime and public transport. In M. Tonry (Ed.), *Crime and justice: An annual review of research* (Vol. 27, pp. 169–233). Chicago: University of Chicago Press.

Clarke, R. V., & Weisburd, D. (1994). Diffusion of crime control benefits: Observations on the reverse of displacement. In R. V. Clarke (Ed.), *Crime prevention studies* (Vol. 2, pp. 165–183). Monsey, NY: Criminal Justice Press.

Cogan, M. B. (1996, May). Diagnosis and treatment of bipolar disorder in adolescents and children. *Psychiatric Times, 13*(5) [Online]. Available at http://www.mhsource.com

Cohen, B. (1980). *Deviant street networks: Prostitution in New York City.* Lexington, MA: Lexington Books.

Cohen, J., Gorr, W. L., & Olligschlaeger, A. M. (2007). Leading indicators and spatial interactions: A crime-forecasting model for proactive police deployment. *Geographical Analysis, 39*(1), 105–127.

Cohen, L. E., & Felson, M. (1979). Social change and crime rate trends: A routine activity approach. *American Sociological Review, 44,* 588–608.

Colquhoun, P. (1969). *A treatise on the police of the metropolis.* Montclair, NJ: Patterson Smith. (Original work published 1795)

Conklin, J. (1994). *Art crime.* Westport, CT: Praeger.

Cornish, D., & Clarke, R. V. (Eds.). (1986). *The reasoning criminal.* New York: Springer-Verlag.

Cornish, D. B., & Clarke, R. V. (2003). Opportunities, precipitators, and criminal decisions: A reply to Wortley's critique of situational crime prevention. In M. J. Smith & D. B. Cornish (Eds.), *Theory for practice in situational crime prevention* (pp. 41–96). Monsey, NY: Criminal Justice Press.

Crofts, P. (2003). White collar punters: Stealing from the boss to gamble. *Current Issues in Criminal Justice, 15*(1), 40–52.

Cromwell, P. (2003). *In their own words: Criminals on crime: An anthology.* Los Angeles: Roxbury.

Cromwell, P. F., Olson, J. N., & Avary, D. W. (1991). *Breaking and entering: An ethnographic analysis of burglary.* Newbury Park, CA: Sage.

Cromwell, P. F., Olson, J. N., & Avary, D. W. (1993). Who buys stolen property: A new look at criminal receiving. *Journal of Crime and Justice, 16*(1), 75–95.

Crowe, T. D. (1990, Fall). Designing safer schools. *School Safety, 9,* 9–13.

Crowe, T. D. (1991). *Crime prevention through environmental design: Applications of architectural design and space management concepts.* Boston: Butterworth-Heinemann.

Crowe, T. D., & Zahm, D. (1994). Crime prevention through environmental design. *Land Management, 7*(1), 22–27.

Cusson, M. (1990). *Croissance et decroissance du crime* [The rise and fall of crime]. Paris: Presses Universitaires de France.

Cusson, M. (1993). A strategic analysis of crime: Criminal tactics as responses to precriminal situations. In R. V. Clarke & M. Felson (Eds.), *Routine activity and rational choice: Advances in criminological theory* (Vol. 5, pp. 295–304). New Brunswick, NJ: Transaction Books.

Dabney, D. A., Hollinger, R. C., & Dugan, L. (2004). Who actually steals? A study of covertly observed shoplifters. *Justice Quarterly, 21*(4), 693–728.

David, F. (2000). *Human smuggling and trafficking: An overview of the response at the federal level* (Paper No. 24, Research and Public Policy Series). Canberra: Australian Institute of Criminology.

Davis, R. C., Maxwell, C. D., & Taylor, B. (2006). Preventing repeat incidents of family violence: Analysis of data from three field experiments. *Journal of Experimental Criminology, 2*(2), 183–210.

Davis, R., & Pease, K. (2000). Crime, technology and the future. *Security Journal, 13*(2), 59–64.

Decker, D., Shichor, D., & O'Brien, R. (1982). *Urban structure and victimization.* Lexington, MA: Lexington Books.

Dedel, J. K. (2005). *School vandalism and break-ins.* Washington, DC: U.S. Department of Justice, Office of Community Oriented Policing Services.

DiLonardo, R. L. (1997). The economic benefit of electronic article surveillance. In R. V. Clarke (Ed.), *Situational crime prevention: Successful case studies* (2nd ed., pp. 122–131). New York: Harrow & Heston.

Dinkes, R., Cataldi, E. F., & Lin-Kelly, W. (2007). *Indicators of school crime and safety* (NCES 2008-021/NCJ219553). Washington, DC: U.S. Department of Education, National Center for Education Statistics, Institute of Education Science; and U.S. Department of Justice, Bureau of Justice Statistics, Office of Justice Programs.

Dolmen, L. (Ed.). (1990). *Crime trends in Sweden, 1988.* Stockholm: National Council for Crime Prevention.

Duffala, D. C. (1976). Convenience stores, armed robbery, and physical environmental features. *American Behavioral Scientist, 20*(2), 227–245.

Eck, J. E. (1995). A general model of the geography of illicit retail marketplaces. In J. E. Eck & D. Weisburd (Eds.), *Crime and place* (pp. 67–93). Monsey, NY: Criminal Justice Press.

Eck, J. E. (1997). Preventing crime at places. In L. W. Sherman, D. Gottfredson, D. MacKenzie, J. E. Eck, P. Reuter, & S. Bushway (Eds.), *Preventing crime: What works, what doesn't, what's promising.* Washington, DC: National Institute of Justice.

Eck, J., Chainey, S., Cameron, J. G., Leitner, M., & Wilson, R. E. (2005). *Mapping crime: Understanding hot spots.* Washington, DC: National Institute of Justice.

Eck, J., & Weisburd, D. (Eds.). (1995). *Crime and place.* Monsey, NY: Criminal Justice Press.

Edmunds, M., Hough, M., & Urquia, N. (1996). *Tackling local drug markets* (Paper No. 80, Crime Detection and Prevention Series). London: British Home Office Research Publications.

Ekblom, P. (1997). Gearing up against crime: A dynamic framework to help designers keep up with the adaptive criminal in a changing world. *International Journal of Risk Security and Crime Prevention, 2*(4), 249–265.

Ekblom, P. (1999). Can we make crime prevention adaptive by learning from other evolutionary struggles? *Studies on Crime & Crime Prevention, 8*(1), 27–51.

Ekblom, P. (2002). Future imperfect: Preparing for the crimes to come. *Criminal Justice Matters, 46*(Winter), 38–40. London: Kings College, Centre for Crime and Justice Studies.

Ekblom, P. (2005a). Designing products against crime. In N. Tilley (Ed.), *Handbook of crime prevention and community safety* (pp. 203–244). Cullompton, UK: Willan.

Ekblom, P. (2005b). How to police the future: Scanning for scientific and technological innovations which generate potential threats and opportunities in crime, policing and crime reduction. In M. Smith & N. Tilley (Eds.), *Crime science: New approaches to preventing and detecting crime* (pp. 27–55). Cullompton, UK: Willan.

Ekblom, P., & Tilley, N. (2000). Going equipped: Criminology, situational crime prevention, and the resourceful offender. *British Journal of Criminology, 40*(3), 376–398.

Ellickson, R. C. (1996). Controlling chronic misconduct in city spaces: Of panhandlers, skid rows, and public-space zoning. *Yale Law Journal, 105,* 1165–1248.

Ellickson, R. C., & Been, V. I. (2005). *Land use controls: Cases and materials* (3rd ed.). New York: Aspen Law & Business Publications.

Ellington, D. (1976). *Music is my mistress.* New York: De Capo Press.

Epstein, J. (1997). *Fixing broken barroom windows. Prevention file, tobacco and other drugs* (No. 12, pp. 1–6). Newton, MA: Higher Education Center for Alcohol and Other Drug Prevention. Available: http://www.edc.org/hec/pubs/articles/fixwindows.html

Espelage, D. L., & Swearer, S. M. (Eds.). (2004). *Bullying in American schools: A social-ecological perspective on prevention and intervention.* Mahwah, NJ: Lawrence Erlbaum.

Fagan, J., & Freeman, R. B. (1999). Crime and work. In M. Tonry (Ed.), *Crime and justice: A review of research* (Vol. 25, pp. 225–290). Chicago: University of Chicago Press.

Farrell, G. (1995). Preventing repeat victimization. In M. Tonry & D. Farrington (Eds.), *Crime and justice: A review of research* (Vol. 19, pp. 469–534). Chicago: University of Chicago Press.

Farrell, G. (1998). Routine activities and drug trafficking: The case of the Netherlands. *International Journal of Drug Policy, 9*(1), 21–32.

Farrell, G. (2001). Crime prevention. In C. D. Bryant (Ed.), *Encyclopedia of criminology & deviant behavior* (pp. 124–133). London: Taylor & Francis.

Farrell, G., & Pease, K. (1993). *Once bitten, twice bitten: Repeat victimization and its implications for crime prevention* (Crime Prevention Unit Paper 46.) London: Home Office.

Farrell, G., & Pease, K. (2001). *Repeat victimization.* Monsey, NY: Criminal Justice Press.

Farrell, G., Tseloni, A., & Pease, K. (2005). Repeat victimization in the ICVS and the NCVS. *Crime Prevention and Community Safety: An International Journal, 7*(3), 7–18.

Fattah, E. A. (1991). *Understanding criminal victimization: An introduction to theoretical victimology.* Scarborough, Ontario, Canada: Prentice Hall.

Federal Bureau of Investigation. (2008a). *Law enforcement officers killed and assaulted.* Washington, DC: Government Printing Office.

Federal Bureau of Investigation. (2008b). *Uniform Crime Reports: Crime in the United States.* Washington, DC: Government Printing Office.

Federal Trade Commission. (2009). *Fighting fraud with the Red Flags Rule: How-to guide for business.* Washington, DC: Author.

Feeney, F. (1986). Robbers as decision makers. In D. Cornish & R. V. Clarke (Eds.), *The reasoning criminal* (pp. 53–71). New York: Springer-Verlag.

Fein, R. A., Vossekuil, B., & Pollack, W. S. (2002). *Threat assessment in schools: A guide to managing threatening situations and to creating safe school climates.* Washington, DC: U.S. Secret Service and U.S. Department of Education.

Felson, M. (1987). Routine activities and crime prevention in the developing metropolis. *Criminology, 25,* 911–931.

Felson, M. (1995a). How buildings can protect themselves against crime. *Lusk Review for Real Estate Development and Urban Transformation, 1*(1), 1–7.

Felson, M. (1995b). Those who discourage crime. In J. E. Eck & D. Weisburd (Eds.), *Crime and place* (pp. 53–66). Monsey, NY: Criminal Justice Press.

Felson, M. (1997). Technology, business and crime. In M. Felson & R. V. Clarke (Eds.), *Business and crime prevention* (pp. 81–96). Monsey, NY: Willow Tree Press.

Felson, M. (1998). *Crime and everyday life* (2nd ed.). Thousand Oaks, CA: Pine Forge.

Felson, M. (2003). The process of co-offending. In M. Smith & D. Cornish (Eds.), *Theory for practice in situational crime prevention* (pp. 149–167). Monsey, NY: Criminal Justice Press.

Felson, M. (2006). *Crime and nature.* Thousand Oaks, CA: Sage.

Felson, M. (2009). The natural history of extended co-offending. *Trends in Organized Crime, 12*(2), 159–165.

Felson, M., Belanger, M. E., Bichler, G. M., Bruzinski, C. D., Campbell, G. S., Fried, C. L., Grofik, K. C., Mazur, I. S., O'Regan, A. B., Sweeney, P. J., Ullman, A. L., & Williams, L. M. (1996). Redesigning hell: Preventing crime and disorder at the Port Authority Bus Terminal. In R. V. Clarke (Ed.), *Preventing mass transit crime* (pp. 5–92). Monsey, NY: Criminal Justice Press.

Felson, M., Berends, R., Richardson, B., & Veno, A. (1997). Reducing pub hopping and related crime. In R. Homel (Ed.), *Reducing crime, public intoxication and injury* (pp. 115–132). Monsey, NY: Criminal Justice Press.

Felson, M., & Clarke, R. V. (1995). Routine precautions, criminology, and crime prevention. In H. D. Barlow (Ed.), *Crime and public policy: Putting theory to work* (pp. 179–190). Boulder, CO: Westview.

Felson, M., & Clarke, R. V. (1997a). *Business and crime prevention.* Monsey, NY: Willow Tree Press.

Felson, M., & Clarke, R. V. (1997b). The ethics of situational crime prevention. In G. Newman, R. V. Clarke, & S. G. Shoham (Eds.), *Rational choice and situational crime prevention: Theoretical foundations* (pp. 197–218). Dartmouth, UK: Ashgate.

Felson, M., & Clarke, R. V. (1998). *Opportunity makes the thief* (Paper No. 98, Police Research Series). London: British Home Office Research Publications.

Felson, M., & Cohen, L. E. (1980). Human ecology and crime: A routine activity approach. *Human Ecology, 8,* 389–406.

Felson, M., & Cohen, L. E. (1981). Modeling crime rate trends—A criminal opportunity perspective. *Journal of Research in Crime and Delinquency, 18,* 138–164 (as corrected, 1982, *19,* 1).

Felson, M., Dickman, D., Glenn, D., Kelly, L., Lambard, L., Maher, L., Nelson-Green, L., Ortega, C., Preiser, T., Rajendran, A., Ross, T., Tous, L., & Veil, J. (1990). Preventing crime at Newark's subway stations. *Security Journal, 1*(3), 137–142.

Felson, M., & Gottfredson, M. (1984). Adolescent activities near peers and parents. *Journal of Marriage and the Family, 46,* 709–714.

Felson, M., & Peiser, R. (Eds.). (1998). *Reducing crime through real estate development and management.* Washington, DC: Urban Land Institute.

Felson, R. B. (1996). Big people hit little people: Sex differences in physical power and interpersonal violence. *Criminology, 34,* 433–452.

Felson, R. B. (2001). Blame analysis: Accounting for the behavior of protected groups. In S. Cole (Ed.), *What's wrong with sociology.* New Brunswick, NJ: Transaction Publications.

Felson, R. B. (2002). *Violence and gender reexamined.* Washington, DC: American Psychological Association.

Felson, R. B., Ackerman, J., & Yeon, S. J. (2003). The infrequency of family violence. *Journal of Marriage and Family, 65,* 622–634.

Felson, R. B., & Felson, S. (1993, September/October). Predicaments of men and women. *Society,* pp. 2016–2020.

Felson, R. B., & Haynie, D. L. (2002). Pubertal development, social factors, and delinquency among adolescent boys. *Criminology, 40*(4), 967–988.

Felson, R. B., & Messner, S. F. (1996). To kill or not to kill? Lethal outcomes in injurious attacks. *Criminology, 34,* 519–546.

Felson, R. B., & Outlaw, M. C. (2007). The control motive and marital violence. *Violence and Victims, 22*(4), 387–407.

Finkelhor, D., Mitchell, K. J., & Wolak, J. (2000). *Online victimization: A report on the nation's youth.* Alexandria, VA: National Center for Missing and Exploited Children.

Finney, A., Wilson, D., Levi, M., Sutton, M., & Forest, S. (2005). *Handling stolen goods: Findings from the 2002–03 British Crime Survey and 2003 Offending, Crime and Justice Survey.* London: Home Office Online Report 38/05. Home Office Research, Development and Statistics Directorate.

Fisher, B. S., Cullen, F. T., & Turner, M. G. (2000). *The sexual victimization of college women* (Research Report, National Institute of Justice & Bureau of Justice Statistics). Washington, DC: U.S. Department of Justice.

Fisher, B. S., Daigle, L. E., & Cullen, F. T. (2003). Reporting sexual victimization to the police and others: Results from a national-level study of college women. *Criminal Justice and Behavior, 30*(1), 6–38.

Fisher, B. S., Daigle, L. E., & Cullen, F. T. (2007). Assessing the efficacy of the protective action-completion nexus for sexual victimizations. *Violence and Victims, 22*(1), 18–42.

Fisher, B., & Nasar, J. L. (1995). Fear spots in relation to microlevel physical cues: Exploring the overlooked. *Journal of Research in Crime and Delinquency, 32,* 214–239.

Fisher, B. S., & Sloan, J. J. (1995). *Campus crime: Legal, social, and policy perspectives.* Springfield, IL: Charles C Thomas.

Fisher, B. S., Sloan, J. J., Cullen, F. T., & Lu, C. (1997). The on-campus victimization patterns of students: Implications for crime prevention by students in post-secondary institutions. In S. P. Lab (Ed.), *Crime prevention at a crossroads* (pp. 101–126). Cincinnati, OH: Anderson.

Fisher, D. A. (1999). *Environmental strategies to prevent alcohol problems on college campuses.* Washington, DC: Office of Juvenile Justice and Delinquency Prevention.

Friedrichs, D. O. (2006). *Trusted criminals: White collar crime in contemporary society* (3rd ed.). Belmont, CA: Wadsworth.

Gabor, T. (1994). *Everybody does it! Crime by the public.* Toronto: University of Toronto Press.

Galovski, T. E., & Blanchard, E. B. (2004). Road rage: A domain for psychological intervention? *Aggression and Violent Behavior: A Review Journal, 9*(3), 105–127.

Gans, H. (1962). *The urban villagers.* New York: Free Press.

Garase, M. L. (2006). *Road rage.* New York: LFB Scholarly.

Garbarino, J. (1978). The human ecology of school crime: A case for small schools. In E. Wenk & N. Harlow (Eds.), *School crime and disruption: Prevention models* (pp. 123–133). Washington, DC: National Institute of Education.

Gardiner, R. A. (1978). *Design for safe neighborhoods.* Washington, DC: Law Enforcement Assistance Administration.

Geis, G., & Meier, R. F. (1977). *White-collar crime: Classic and contemporary views.* New York: Free Press.

George, B., & Button, M. (2000). *Private security.* Leicester, UK: Perpetuity Press.

Gilfillan, S. C. (1935). *The sociology of invention.* Cambridge: MIT Press.

Gill, M. (Ed.). (1994). *Crime at work: Studies in security and crime prevention* (Vol. 1). Leicester, UK: Perpetuity Press.

Gill, M. (Ed.). (1998). *Crime at work: Studies in security and crime prevention* (Vol. 2). Leicester, UK: Perpetuity Press.

Gill, M. (2000a). *Commercial robbery: Offenders' perspectives on security and crime prevention.* London: Blackstone.

Gill, M. (2000b). *Insurance fraud: Causes, patterns, and prevention.* Unpublished doctoral dissertation, University of Leicester, England.

Glensor, R. W., & Peak, K. J. (2004). *Cruising.* Washington, DC: Office of Community Oriented Policing Services.

Goldstein, H. (1990). *Problem-oriented policing.* New York: McGraw-Hill.

Goldstein, H. (1997). The new policing: Confronting complexity. In P. F. Cromwell & R. G. Dunham (Eds.), *Crime and justice in America: Realities and future prospects* (pp. 95–103). New York: Prentice Hall.

Goldstein, H. (2003). On further developing problem-oriented policing: The most critical need, the major impediments, and a proposal. In J. Knutsson (Ed.), *Problem-oriented policing: From innovation to mainstream* (pp. 13–47). Monsey, NY: Criminal Justice Press.

Goldstein, H. (2005). *Problem oriented policing.* Interview with Herman Goldstein. Available: http://www.popcenter.org/learning/goldstein_interview/

Gottfredson, D. C. (1985). Youth employment, crime and schooling: A longitudinal study of a national sample. *Developmental Psychology, 21,* 419–432.

Gottfredson, D. C. (2001). *Schools and delinquency.* Cambridge, UK: Cambridge University Press.

Gottfredson, D. C., Gerstenblith, S. A., Soulé, D. A., Womer, S. C., & Lu, S. (2004). Do after school programs reduce delinquency? *Prevention Science, 5,* 253–266.

Gottfredson, G. D., & Gottfredson, D. C. (1985). *Victimization in schools.* New York: Plenum.

Gottfredson, M. (1984). *Victims of crime: The dimensions of risk* (Report No. 81, Home Office Research and Planning Unit). London: British Home Office Research Publications.

Gottfredson, M., & Hirschi, T. (1990). *A general theory of crime.* Stanford, CA: Stanford University Press.

Graham, K., & Homel, R. (2008). *Raising the bar: Preventing aggression in and around bars, pubs and clubs.* Monsey, NY: Criminal Justice Press, and London: Willan Press.

Green, L. (1996). *Policing places with drug problems.* Thousand Oaks, CA: Sage.

Greenberg, J. (2002). Who stole the money, and when? Individual and situational determinants of employee theft. *Organizational Behavior and Human Decision Processes, 89*(1), 985–1003.

Greenberger, E., & Steinberg, L. (1986). *When teenagers work: The psychological and social costs of adolescent employment.* New York: Basic Books.

Gregory, T. (2000). *School reform and the no-man's-land of high school size.* Available: http://www.smallschoolsproject.org/PDFS/gregory.pdf

Greif, J. L., & Furlong, M. J. (2006). The assessment of school bullying: Using theory to inform practice. *Journal of School Violence, 5*(3), 33–50.

Grey, M. A., & Anderson-Ryan, W. (1994). Serving & scamming: A qualitative study of employee theft in one chain restaurant. *Security Journal, 5*(4), 200–211.

Groff, E. R. (2007). Simulation for theory testing and experimentation: An example using routine activity theory and street robbery. *Journal of Quantitative Criminology, 23*(2), 75–103.

Grogan, P., & Proscio, T. (2001). *Comeback cities: A blueprint for urban neighborhood revival.* Boulder, CO: Westview.

Grubesic, T. H., & Mack, E. A. (2008). Spatio-temporal interaction of urban crime. *Journal of Quantitative Criminology, 24*(3), 285–306.

Guerrette, R. T., & Clarke, R. V. (2003). Product life cycles and crime: Automated teller machines and robbery. *Security Journal, 16*(1), 7–18.

Hagan, J., & McCarthy, B. (1998). *Mean streets: Youth crime and homeless.* New York: Cambridge University Press.

Hakim, S., Rengert, G. F., & Shachmurove, Y. (2001). Target search of burglars: A revised economic model. *Papers in Regional Science, 80,* 121–137.

Hardin, G. (1968). The tragedy of the commons. *Science, 162,* 1243–1248.

Harding, R., Morgan, F. H., Indermaur, D., Ferrante, A. M., & Blagg, H. (1998). Road rage and the epidemiology of violence: Something old, something new. *Studies on Crime and Crime Prevention, 7*(2), 221–238.

Harries, K. (1999). *Mapping crime: Principle and practice.* Washington, DC: National Institute of Justice.

Hartshorne, H., & May, M. A. (1928–1930). *Studies in the nature of character.* New York: Macmillan.

Hawley, A. H. (1971). *Urban society: An ecological approach.* New York: Ronald Press.

Hayes, R. (1997a). Retail crime control: A new operational strategy. *Security Journal, 8*(3), 225–232.

Hayes, R. (1997b). Retail theft: An analysis of apprehended shoplifters. *Security Journal, 8*(3), 233–246.

Hayes, R. (1999). Shop theft: An analysis of shoplifter perceptions and situational factors. *Security Journal, 12*(2), 7–18.

Hayes, R. (2000). US retail store detectives: An analysis of their focus, selection and training. *Security Journal, 13*(1), 7–20.

Hayes, R. (2003). Loss prevention: Senior management views on current trends and issues. *Security Journal, 16*(2), 7–20.

Hearnden, I., & Magill, C. (Eds.). (2004). *Decision-making by house burglars: Offenders' perspectives* (Vol. 259). London: Home Office Research, Development and Statistics Directorate.

Henry, B., Caspi, A., & Moffitt, T. E. (1999). Staying in school protects boys with poor self-regulation in childhood from later crime: A longitudinal study. *International Journal of Behavioral Development, 23*(4), 1049–1073.

Hesseling, R. B. (1993). Displacement: A review of the empirical literature. In R. V. Clarke (Ed.), *Crime prevention studies* (Vol. 3, pp. 197–270). Monsey, NY: Criminal Justice Press.

Hindelang, M., Gottfredson, M., & Garafolo, J. (1978). *Victims of personal crime: An empirical foundation for a theory of personal victimization.* Cambridge, MA: Ballinger.

Hirschi, T. (1969). *Causes of delinquency.* Berkeley: University of California Press.

Hirschi, T., & Gottfredson, M. (1993a). Commentary: Testing the general theory of crime. *Journal of Research in Crime and Delinquency, 30*(1), 47–54.

Hirschi, T., & Gottfredson, M. R. (1993b). Control theory and life-course perspective. *Studies in Crime Prevention, 4*(2), 131–142.

Hirschi, T., & Gottfredson, M. R. (Eds.). (1994). *The generality of deviance.* New Brunswick, NJ: Transaction.

Hirschi, T., & Gottfredson, M. R. (2000). In defense of self-control. *Theoretical Criminology, 4*(1), 55–69.

Hirschi, T., & Stark, R. (1969). Hellfire and delinquency. *Social Problems, 17,* 202–212.

Ho, T. N. (1998). Prevention of art theft at commercial art galleries. *Studies on Crime and Crime Prevention, 7*(2), 213–219.

Hollinger, R. C. (1993). *National retail security survey.* Gainesville: University of Florida, Department of Sociology.

Hollinger, R. C. (1997). Measuring crime and its impact in the business environment. In M. Felson & R. V. Clarke (Eds.), *Business and crime prevention* (pp. 57–79). Monsey, NY: Criminal Justice Press.

Hollinger, R. C., Langton, L., & Adams, A. (2007). *Organized retail crime and e-fencing.* Paper presented at the annual meeting of the American Society of Criminology, Atlanta, GA.

Holzman, H. R. (1996). Criminological research on public housing: Toward a better understanding of people, places, and spaces. *Crime and Delinquency, 42,* 361–378.

Homel, R. (Ed.). (1997). *Policing for prevention: Reducing crime, public intoxication, and injury.* Monsey, NY: Criminal Justice Press.

Homel, R., Hauritz, M., McIlwain, G., Wortley, R., & Carvolth, R. (1997). Preventing drunkenness and violence around nightclubs in a tourist resort. In R. V. Clarke (Ed.), *Situational crime prevention: Successful case studies* (2nd ed., pp. 263–282). New York: Harrow & Heston.

Hope, T. J. (1982). *Burglary in schools: The prospects for prevention* (Paper No. 11, Research and Planning Unit). London: British Home Office Research Publications.

Hope, T. (2007). The distribution of household property crime victimization: Insights from the British Crime Survey. In M. Hough & M. Maxfield (Eds.), *Surveying crime in the 21st century* (pp. 99–124). Monsey, NY: Criminal Justice Press.

Hunt, A. (1999). The purity wars: Making sense of moral militancy. *Theoretical Criminology, 3*(4), 409–436.

Hunter, R. D., & Jeffery, C. R. (1997). Preventing convenience store robbery through environmental design. In R. V. Clarke (Ed.), *Situational crime prevention: Successful case studies* (2nd ed., pp. 191–199). New York: Harrow & Heston.

Indermaur, D. (1995). *Violent property crime.* Sydney: Federation Press.

Jacobs, B. A., & Wright, R. T. (1999). Stick-up, street culture, and offender motivation. *Criminology, 37*(1), 149–173.

Jacobs, J. (1961). *Death and life of great American cities.* New York: Random House.

Jacobson, J. (1999). *Policing drug hot-spots* (Paper No. 109, Police Research Series). London: British Home Office Research Publications.

James, M. (2000). *Art crime* (Trends and Issues in Crime and Criminal Justice, No. 170). Available: http://203.221.207.15/publications/tandi/ti170.pdf

Jansen, A. C. (1995). The development of a "legal" consumers' market for cannabis: The "coffee shop" phenomenon. In E. Leuw & I. H. Marshall (Eds.), *Between prohibition and legalization: The Dutch experiment in drug policy* (pp. 169–181). Amsterdam: Kugler.

Jeffery, C. R. (1971). *Crime prevention through environmental design.* Beverly Hills, CA: Sage.

Johns, T., & Hayes, R. (2003). Behind the fence: Buying and selling stolen merchandise. *Security Journal, 16*(4), 29–44.

Johnson, B., Goldstein, P., Preble, E., Schmeidler, J., Lipton, D. S., Spunt, B., & Miller, T. (1985). *Taking care of business: The economics of crime by heroin abusers.* Lexington, MA: Lexington Books.

Johnson, S. D., & Bowers, K. J. (2004a). The burglary as clue to the future: The beginnings of prospective hot-spotting. *European Journal of Criminology, 1,* 237–255.

Johnson, S. D. & Bowers, K. J. (2004b). The stability of space-time clusters of burglary. *British Journal of Criminology, 44,* 55–65.

Johnson, S. D., & Bowers, K. J. (2007). Burglary prediction: The role of theory, flow and friction. In G. Farrell, K. J. Bowers, S. D. Johnson, & M. Townsley (Eds.), *Imagination for crime prevention: Essays in honor of Ken Pease* (pp. 203–224). Monsey, NY: Criminal Justice Press.

Johnston, L. D., O'Malley, P. M., Bachman, J. G., & Schulenberg, J. E. (2007). *Monitoring the Future national results on adolescent drug use: Overview of key findings, 2006.* (NIH Publication No. 07–6202). Bethesda, MD: National Institute on Drug Abuse.

Junger, M., & Marshall, I. H. (1997). The interethnic generalizability of social control theory: An empirical test. *Journal of Research in Crime and Delinquency, 34*(1), 79–112.

Junger-Tas, J., & van Kesteren, J. N. (1999). *Bullying and delinquency in a Dutch school population.* The Hague, The Netherlands: Kugler.

Kelling, G. L. (1999). *Broken windows and police discretion* (Report No. 178259). Washington, DC: National Institute of Justice.

Kelling, G. L., & Coles, C. (1996). *Fixing broken windows: Restoring order and reducing crime in our communities.* New York: Free Press.

Kelling, G. L., Pate, T., Dieckman, D., & Brown, C. (1974). *The Kansas City preventive patrol experiment: A summary report.* Washington, DC: Police Foundation.

Kennedy, L. W., & Forde, D. R. (1990). Routine activities and crime: An analysis of victimization in Canada. *Criminology, 28,* 101–115.

Kennedy, L. W., & Forde, D. R. (1999). *When push comes to shove: A routine conflict approach to violence.* Albany: State University of New York Press.

Kepenekci, Y. K., & Cinkir, S. (2006). Bullying among Turkish high school students. *Child Abuse and Neglect, 30*(2), 193–204.

Kidwell, R. E., & Martin, C. L. (2005). *Managing organizational deviance.* Thousand Oaks, CA: Sage.

King, M., & Brearley, N. (1996). *Public order policing: Contemporary perspectives on strategy and tactics.* Leicester, UK: Perpetuity Press.

Klein, M. W. (1971). *Street gangs and street workers.* Englewood Cliffs, NJ: Prentice Hall.

Klein, M. W. (1995). *The American street gang: Its nature, prevalence, and control.* New York: Oxford University Press.

Klein, M. W., Maxson, C. L., & Cunningham, L. C. (1991). "Crack," street gangs, and violence. *Criminology, 29,* 623–650.

Klockars, C. B. (1974). *The professional fence.* New York: Free Press.

Knutsson, J. (2000). Swedish drug markets and drug policy. In M. Natarajan & M. Hough (Eds.), *Illegal drug markets: From research to prevention policy* (pp. 179–201). Monsey, NY: Criminal Justice Press.

Knutsson, J., & Clarke, R. V. (Eds.). (2006). *Putting theory to work: Implementing situational prevention and problem-oriented policing.* Monsey, NY: Criminal Justice Press.

Knutsson, J., & Kuhlhorn, E. (1997). Macro measures against crime: The example of check forgeries. In R. V. Clarke (Ed.), *Situational crime prevention: Successful case studies* (2nd ed., pp. 113–121). New York: Harrow & Heston.

Kuenstle, M. W., Clark, N. M., & Schneider, R. (Eds.). (2003). *Florida safe school design guidelines: Strategies to enhance security and reduce vandalism.* Tallahassee: Florida Department of Education, Office of Educational Facilities.

Kuhlhorn, E. (1997). Housing allowances in a welfare society: Reducing the temptation to cheat. In R. V. Clarke (Ed.), *Situational crime prevention: Successful case studies* (2nd ed., pp. 235–241). New York: Harrow & Heston.

Lab, S. P., & Clark, R. D. (1997). Crime prevention in schools: Individual and collective responses. In S. P. Lab (Ed.), *Crime prevention at a crossroads* (pp. 127–140). Cincinnati, OH: Anderson.

Langton, L., & Hollinger, R. C. (2005). Correlates of crime losses in the retail industry. *Security Journal, 18*(3), 27–44.

Langworthy, R. (1989). Do stings control crime? An evaluation of a police fencing operation. *Justice Quarterly, 6*(1), 27–45.

Langworthy, R., & LeBeau, J. (1992). The spatial evolution of a sting clientele. *Journal of Criminal Justice, 20*(2), 135–146.

Lasley, J. (1998). *"Designing out" gang homicides and street assaults: Research in brief.* Washington, DC: National Institute of Justice.

Lasley, J. R., & Rosenbaum, J. L. (1988). Routine activities and multiple personal victimization. *Sociology and Social Research, 73*(1), 47–50.

LaVigne, N. (1996). Safe transport: Security by design on the Washington Metro. In R. V. Clarke (Ed.), *Preventing mass transit crime* (pp. 163–197). Monsey, NY: Criminal Justice Press.

LaVigne, N. (1997). Security by design on the Washington Metro. In R. V. Clarke (Ed.), *Situational crime prevention: Successful case studies* (2nd ed., pp. 283–299). New York: Harrow & Heston.

Lawton, B. A., Taylor, R. B., & Luongo, A. J. (2005). Police officers on drug corners in Philadelphia, drug crime and violent crime: Intended, diffusion and displacement impacts. *Justice Quarterly, 22*(4), 427–451.

LeBeau, J. (1987). The journey to rape: Geographic distances and the rapist's method of approaching the victim. *Journal of Police Sciences and Administration, 15*(2), 129–136.

LeBeau, J., & Langworthy, R. (1992). The spatial distribution of sting targets. *Journal of Criminal Justice, 20,* 541–551.

Locker, J. P., & Godfrey, B. (2006). Ontological boundaries and temporal watersheds in the development of white-collar crime. *British Journal of Criminology, 46*(6), 976–992.

Macintyre, S., & Homel, R. (1997). Danger on the dance floor: A study of interior design, crowding and aggression in nightclubs. In R. Homel (Ed.), *Policing for prevention: Reducing crime, public intoxication and injury* (pp. 91–113). Monsey, NY: Criminal Justice Press.

Madsen, T., & Eck, J. (2008). *Spectator violence in stadiums.* Washington, DC: U.S. Department of Justice, Office of Community Oriented Policing Services.

Mannon, J. M. (1997). Domestic and intimate violence: An application of routine activities theory. *Aggression and Violent Behavior, 2*(1), 9–24.

Masuda, B. (1993). Credit card fraud prevention: A successful retail strategy. In R. V. Clarke (Ed.), *Crime prevention studies* (Vol. 1, pp. 121–134). Monsey, NY: Criminal Justice Press.

Masuda, B. (1997). Reduction of employee theft in a retail environment: Displacement vs. diffusion of benefits. In R. V. Clarke (Ed.), *Situational crime prevention: Successful case studies* (2nd ed., pp. 183–190). New York: Harrow & Heston.

Matthews, R. (2002). *Armed robbery.* Cullompton, Devon, UK: Willan.

Maxfield, M. G. (1989). Circumstances in supplementary homicide reports: Variety and validity. *Criminology, 27,* 671–695.

Maxfield, M. G., & Clarke, R. V. (Eds.). (2004). *Understanding and preventing car theft.* Monsey, NY: Criminal Justice Press.

May, T., Hough, M., & Edmunds, M. (2000). Sex markets and drug markets: Examining the links. *Crime Prevention and Community Safety, 2*(1), 25–41.

Mayhew, C. (2000). *Violence in the workplace—Preventing armed robbery: A practical handbook* (Public Policy Series, No. 33). Canberra: Australian Institute of Criminology.

Mayhew, P., Clarke, R. V., Burrows, J. N., Hough, J. M., & Winchester, S. W. (1979). *Crime in public view* (No. 2049, Home Office Research Study). London: British Home Office Research Publications.

Mayhew, P., Clarke, R. V., & Eliot, D. (1989). Motorcycle theft, helmet legislation, and displacement. *Howard Journal of Criminal Justice, 28*(1), 1–8.

Mayhew, P., Clarke, R. V., Sturman, A., & Hough, J. M. (1976). *Crime as opportunity.* London: British Home Office Research Publications.

McCarthy, B., Felmlee, D., & Hagan, J. (2004). Girl friends are better: Gender, friends and crime among school and street youth. *Criminology, 42*(4), 805–835.

McCarthy, B., & Hagan, J. (2001). When crime pays: Capital, competence and criminal success. *Social Forces, 79*(3), 1035–1060.

McCarthy, B., & Hagan, J. (2005). Danger and the decision to offend. *Social Forces, 83*(3), 1065–1096.

McCarthy, B., Hagan, J., & Martin, M. J. (2002). In and out of harm's way: Violent victimization and the social capital of fictive street families. *Criminology, 40*(4), 831–866.

McGarrell, E. F., & Weiss, A. (1996, November). *The impact of increased traffic enforcement on crime.* Paper presented at the annual meeting of the American Society of Criminology, Bloomington, Indiana.

Memphis-Shelby Crime Commission. (2000). *Crime prevention through coordinated and community-based after school programs* (Best Practice Paper No. 6). Memphis, TN: Author.

Meschke, L. L., & Silbereisen, R. K. (1997). The influence of puberty, family processes, and leisure activities on the timing of first sexual experience. *Journal of Adolescence, 20,* 403–418.

Montemayor, R., Adams, G. R., & Gullotta, T. P. (Eds.). (1990). *From childhood to adolescence: A transitional period?* Newbury Park, CA: Sage.

Morrison, S., & O'Donnell, I. (1996). An analysis of the decision making processes of armed robbers. In R. Homel (Ed.), *The politics and practice of situational crime prevention* (pp. 159–188). Monsey, NY: Criminal Justice Press.

Morselli, C., & Tremblay, P. (2004). Criminal achievement, offender networks and the benefits of low self-control. *Criminology, 42*(3), 773–804.

Morselli, C., Tremblay, P., & McCarthy, B. (2006). Mentors and criminal achievement. *Criminology, 44*(1), 17–43.

Nagin, D. S., Farrington, D. P., & Moffitt, T. E. (1995). Life course trajectories of different types of offenders. *Criminology, 33*(1), 111–139.

Natarajan, M. (2000). Understanding the structure of a drug trafficking organization: A conversational analysis. In M. Natarajan & M. Hough (Eds.), *Illegal drug markets: From research to prevention policy* (pp. 273–298). Monsey, NY: Criminal Justice Press.

Natarajan, M. (2006). Understanding the structure of a large heroin distribution network: A quantitative analysis of qualitative data. *Journal of Quantitative Criminology, 22*(2), 171–192.

Natarajan, M., Clarke, R. V., & Johnson, B. D. (1995). Telephones as facilitators of drug dealing: A research agenda. *European Journal of Crime Policy & Research, 3*(3), 137–154.

Natarajan, M., & Hough, M. (Eds.). (2000). *Illegal drug markets: From research to prevention policy.* Monsey, NY: Criminal Justice Press.

New Jersey Legislature. (2005). *School size, violence, achievement and cost.* A report of the Commission on Business Efficiency of the Public Schools. Available: http://www.njleg.state.nj.us/legislativepub/reports/buseff_report.pdf

Newman, G. (2003). *Check and card fraud.* Washington, DC: U.S. Department of Justice, Office of Community Oriented Policing Services.

Newman, G. (2004). *Identity theft.* Washington, DC: U.S. Department of Justice, Office of Community Oriented Policing Services.

Newman, G. (2007). *Sting operations.* Washington, DC: U.S. Department of Justice, Office of Community Oriented Policing Services.

Newman, G., & Clarke, R. V. (2003). *Superhighway robbery: Preventing e-commerce crime.* Cullompton, Devon, UK: Willan.

Newman, O. (1972). *Defensible space: Crime prevention through urban design.* New York: Macmillan.

Ogburn, W. F. (1964). *On culture and social change: Selected papers of William F. Ogburn* (O. D. Duncan, Ed.). Chicago: University of Chicago Press.

O'Kane, J. B., Fisher, R. M., & Green, L. (1994). Mapping campus crime. *Security Journal, 5*(8), 172–180.

Olweus, D. (1978). *Aggression in the schools: Bullies and whipping boys.* New York: Halsted.

Olweus, D. (1993). *Bullying at school: What we know and what we can do.* Oxford, UK: Blackwell.

Organized Crime Task Force, New York State. (1988). *Corruption and racketeering in the New York City construction industry* (New York State School of Industrial and Labor Relations, Cornell University). Ithaca, NY: ILR Press.

Painter, K., & Farrington, D. P. (1997). The crime reducing effect of improved street lighting: The Dudley Project. In R. V. Clarke (Ed.), *Situational crime prevention: Successful case studies* (2nd ed., pp. 209–226). New York: Harrow & Heston.

Painter, K., & Tilley, N. (1999). Editor's introduction: Seeing and being seen to prevent crime. In K. Painter & N. Tilley (Eds.), *Surveillance of public space: CCTV, street lighting and crime prevention* (pp. 1–13). Monsey, NY: Criminal Justice Press.

Patchin, J. W., & Hinduja, S. (2006). Bullies move beyond the schoolyard: A preliminary look at cyberbullying. *Youth Violence and Juvenile Justice, 4,* 148–169.

Paternoster, R., & Simpson, S. (1993). A rational choice theory of corporate crime. In R. V. Clarke & M. Felson (Eds.), *Routine activity and rational choice: Advances in criminological theory* (Vol. 5, pp. 37–58). New Brunswick, NJ: Transaction Books.

Payne, B. K., & Gainey, R. R. (2004). Ancillary consequences of employee theft. *Journal of Criminal Justice, 32*(1), 63–73.

Pease, K. (1992). Preventing burglary on a British public housing estate. In R. V. Clarke (Ed.), *Situational crime prevention: Successful case studies* (pp. 223–229). New York: Harrow & Heston.

Pease, K. (1997). Predicting the future: The roles of routine activity and rational choice theory. In G. Newman, R. V. Clarke, & S. G. Shoham (Eds.), *Rational choice and situational crime prevention: Theoretical foundations* (pp. 233–245). Dartmouth, UK: Ashgate.

Pease, K. (1999). A review of street lighting evaluations: Crime reduction efforts. In K. Painter & N. Tilley (Eds.), *Surveillance of public space: CCTV, street lighting and crime prevention* (pp. 47–76). Monsey, NY: Criminal Justice Press.

Pease, K., Rogerson, M., & Ellingworth, D. (2008). *Future crime trends in the United Kingdom.* Association of British Insurers, General Insurance Research Report No. 7.

Petrosino, A., & Brensilber, D. (2003). The motives, methods, and decision-making of convenience store robberies: Interview with 28 incarcerated offenders in Massachusetts. In M. Smith & D. Cornish (Eds.), *Theory for practice in situational crime prevention* (pp. 237–263). Monsey, NY: Criminal Justice Press.

Phillips, C. (1991). *Multiple victimization and bullying behavior in four schools.* Unpublished manuscript.

Pitts, J., & Smith, P. (1995). *Preventing school bullying* (Paper No. 63, Crime Detection and Prevention Series). London: British Home Office Research Publications.

Pizam, A., & Mansfield, Y. (Eds.). (1996). *Tourism, crime, and international security issues.* Chichester, UK: Wiley.

Potter, G. W., & Kappeler, V. E. (1998). *Constructing crime: Perspectives on making news and social problems.* Prospect Heights, IL: Waveland.

Poyner, B. (1997). An evaluation of walkway demolition on a British housing estate. In R. V. Clarke (Ed.), *Situational crime prevention: Successful case studies* (2nd ed., pp. 59–73). New York: Harrow & Heston.

Poyner, B. (1998). The case for design. In M. Felson & R. B. Peiser (Eds.), *Reducing crime through real estate development and management* (pp. 5–21). Washington, DC: Urban Land Institute.

Poyner, B., & Fawcett, W. H. (1995). *Design for inherent security: Guidance for nonresidential buildings.* London: Construction Industry Research and Information Association.

Poyner, B., & Webb, B. (1991). *Crime free housing.* Oxford, UK: Butterworth.

Pratt, T., & Cullen, F. T. (2000). The empirical status of Gottfredson and Hirschi's general theory of crime: A meta-analysis. *Criminology, 38*(3), 931–964.

Pred, A. (Ed.). (1981). *Space and time in geography: Essays dedicated to Torsten Hagerstrand.* Lund, Sweden: CWK Gleerup.

Randall, P. (1997). *Adult bullies: Perpetrators and victims.* London: Routledge.

Ratcliffe, J. H. (2004). The hotspot matrix: A framework for the spatio-temporal targeting of crime reduction. *Police Practice and Research, 5*(1), 5–23.

Ratcliffe, J. H. (2005). Detecting spatial movement of intra-region crime patterns over time. *Journal of Quantitative Criminology, 21*(1), 103–123.

Ready, J. (2008). *Offender adaptation: Understanding crime displacement from a micro-level perspective.* Unpublished doctoral dissertation, Rutgers University, Newark, NJ.

Rengert, G. (1996). *The geography of illegal drugs.* Boulder, CO: Westview.

Rengert, G., Chakravorty, S., Bole, T., & Henderson, K. (2000). A geographic analysis of illegal drug markets. In M. Natarajan & M. Hough (Eds.). *Illegal drug markets: From research to prevention policy* (pp. 219–239). Monsey, NY: Criminal Justice Press.

Rengert, G. F., Mattson, M. T., & Henderson, K. D. (2001). *Campus security: situational crime prevention in high-density environments.* Monsey, NY: Criminal Justice Press.

Rengert, G. F., & Wasilchick, J. (2000). *Suburban burglary: A tale of two suburbs* (2nd ed.). Springfield, IL: Charles C Thomas.

Reuter, P. (1998). *The mismeasurement of illegal drug markets* (Report RP-613). Santa Monica, CA: RAND.

Rickman, N., & Witt, R. (2003). *The determinants of employee crime in the UK.* London: Center for Economic Policy Research.

Riedel, M. (1999). The decline of arrest and clearance for criminal homicide: Causes, correlates and third parties. *Criminal Justice Policy Review, 9*(3&4), 279–306.

Robinson, J. B. (2008). Measuring the impact of a targeted law enforcement initiative on drug sales. *Journal of Criminal Justice, 36*(1), 90–101.

Roncek, D. W. (1981). Dangerous places: Crime and residential environment. *Social Forces, 60,* 74–96.

Roncek, D. W., & Lobosco, A. (1983). The effect of high schools on crime in their neighborhoods. *Social Science Quarterly, 64,* 598–613.

Roncek, D. W., & Maier, P. A. (1991). Bars, blocks and crimes revisited: Linking the theory of routine activities to the empiricism of "hot spots." *Criminology, 29,* 725–754.

Rosenfeld, R., Bray, T. M., & Egley, A. (1999). Facilitating violence: A comparison of gang-motivated, gang-affiliated, and nongang youth homicide. *Journal of Quantitative Criminology, 15*(4), 495–516.

Ross, H. L. (1992). *Confronting drunk driving: Social policy for saving lives.* New Haven, CT: Yale University Press.

Rossmo, D. K. (1995). Place, space, and police investigations: Hunting serial violent criminals. In J. E. Eck & D. Weisburd (Eds.), *Crime and place* (pp. 217–235). Monsey, NY: Criminal Justice Press.

Rossmo, D. K. (2000). *Geographic profiling.* Boca Raton, FL: CRC Press.

Rossmo, D. K., Davies, A., & Patrick, M. (2004). *Exploring the geo-demographic and distance relationships between stranger rapists and their offences.* London: Home Office, Research, Development and Statistics Directorate.

Runyan, C. W., Bowling, J. M., & Schulman, M. (2005). Potential for violence against teenage retail workers in the United States. *Journal of Adolescent Health, 36*(3), 1–6.

Schneider, J. L. (2003). Prolific burglars and the role of shoplifting. *Security Journal, 16*(2), 49–59.

Schuerman, L. A., & Kobrin, S. (1986). Community careers in crime. In A. Reiss & M. Tonry (Eds.), *Communities and crime* (pp. 67–100). Chicago: University of Chicago Press.

Scott. M. (2001). *Speeding in residential areas.* Washington, DC: U.S. Department of Justice, Office of Community Oriented Policing Services.

Scott, M. (2004). *The benefits and consequences of police crackdowns.* Washington, DC: U.S. Department of Justice, Office of Community Oriented Policing Services.

Scott, M., & Dedel, K. (2006). *Assaults in and around bars* (2nd ed.). Washington, DC: U.S. Department of Justice, Office of Community Oriented Policing Services.

Scott, M., Emerson, N. J., Antonacci, L. B., & Plant, J. B. (2006). *Drunk driving.* Washington, DC: U.S. Department of Justice, Office of Community Oriented Policing Services.

Scott, M., & Goldstein, H. (2005). *Shifting and sharing responsibility for public safety problems.* Washington, DC: U.S. Department of Justice, Office of Community Oriented Policing Services.

Shaw, C., & McKay, H. (1942). *Juvenile delinquency and urban areas.* Chicago: University of Chicago Press.

Shearing, C. D., & Stenning, P. C. (1997). From the Panopticon to Disney World: The development of discipline. In R. V. Clarke (Ed.), *Situational crime prevention: Successful case studies* (2nd ed., pp. 300–304). New York: Harrow & Heston.

Shepherd, J. P., Brickley, M. R., Gallagher, D., & Walker, R. V. (1994). Risk of occupational glass injury in bar staff. *Injury, 25,* 219–220.

Shepherd, J. P., Hugget, R. H., & Kidner, G. (1993). Impact resistance of bar glasses. *Journal of Trauma, 35,* 936–938.

Sherman, C. (2009, February 25). National Retail Federation welcomes legislation to fight organized retail crime, says retailers can't afford losses from theft during recession. News Release. Available at http://www.nrf.com

Sherman, L., Gartin, P. R., & Buerger, M. E. (1989). Hot spots of predatory crime: Routine activities and the criminology of place. *Criminology, 27*(1), 27–56.

Sherman, L., & Weisburd, D. (1995). General deterrent effects of police patrol in crime "hot spots": A randomized controlled trial. *Justice Quarterly, 12,* 625–648.

Simon, H. A. (1957). *Models of man.* New York: Wiley.

Skogan, W. G. (1990). *Disorder and decline: Crime and the spiral of decay in American neighborhoods.* New York: Free Press.

Skogan, W. G. (Ed.). (2006). *Studies in crime and public policy.* New York: Oxford University Press.

Skogan, W. G., & Hartnett, S. M. (1997). *Community policing, Chicago style.* New York: Oxford University Press.

Skogan, W. G., Rosenbaum, D. P., & Hartnett, S. M. (2005). *CLEAR and I-CLEAR: A status report on new information technology and its impact on management, the organization and crime-fighting strategies.* Chicago: Illinois Criminal Justice Information Authority.

Skogan, W. G., Steiner, L., & Benitez, C. (Eds.). (2004). *Community policing in Chicago, year ten: An evaluation of Chicago's alternative policing strategy.* Chicago: Illinois Criminal Justice Information Authority.

Skogan, W., Steiner, L., & Dubois, J. (2002). *Taking stock: Community policing in Chicago.* Washington, DC: U.S. National Institute of Justice.

Sloan, J. H., Kellermann, A. L., Reay, D. T., Ferris, A. J., Rice, C. L., & LoGerfo, J. (1988). Handgun regulations, crime, assaults, and homicide: A tale of two cities. *New England Journal of Medicine, 319,* 1256–1262.

Sloan-Howitt, M., & Kelling, G. D. (1997). Subway graffiti in New York City: "Gettin' up" vs. "meanin' it and cleanin' it." In R. V. Clarke (Ed.), *Situational crime prevention: Successful case studies* (2nd ed., pp. 242–249). New York: Harrow & Heston.

Smigel, E. O., & Ross, H. L. (1970). *Crimes against bureaucracy.* New York: Van Nostrand.

Smith, D. J. (1995). Youth crime and conduct disorders: Trends, patterns and causal explanations. In M. Rutter & D. J. Smith (Eds.), *Psychosocial disorders in youth populations: Time trends and their causes* (pp. 389–489). Chichester, UK: Wiley.

Smith, M. (2005). *Robbery of taxi drivers.* Washington, DC: U.S. Department of Justice, Office of Community Oriented Policing Services.

Smith, M., & Clarke, R. V. (2000). Crime and public transport. In M. Tonry (Ed.). *Crime and justice: A review of research* (Vol. 27, pp. 169–233). Chicago: University of Chicago Press.

Smith, M., & Cornish, D. (Eds.). (2006). *Secure and tranquil travel: Preventing crime and disorder on public transport.* Cullompton, Devon, UK: Willan.

Smith, P. K., Morita, Y., Junger-Tas, J., Olweus, D., Catalano, R., & Slee, P. (Eds.). (1999). *The nature of school bullying: A cross-national perspective.* New York: Routledge.

Smith, P. K., & Sharp, S. (Eds.). (1994). *School bullying: Insights and perspectives.* London: Routledge.

Snyder, H. N., & Sickmund, M. (2006). *Juvenile offenders and victims: 2006 national report* (Prepared for the Office of Juvenile Justice and Delinquency Prevention). Pittsburgh, PA: National Center for Juvenile Justice.

Southall, D., & Ekblom, P. (1985). *Designing for car security: Towards a crime-free car* (Crime Prevention Unit Paper No. 204). London: British Home Office Research Publications.

Sprague, J. R., & Walker, H. M. (2004). *Safe and healthy schools: Practical prevention strategies.* New York: Guilford.

Staff, J., & Uggen, C. (2003). The fruits of good work: Early work experiences and adolescent deviance. *Journal of Research in Crime and Delinquency, 40*(3), 263–290.

Stangeland, P. (1995). *The crime puzzle: Crime patterns and crime displacement in Southern Spain.* Malaga, Spain: Andalusian Inter-University Institute of Criminology (IAIC).

State Education Department. (1994). *A study of safety and security in the public schools of New York State.* Albany: State of New York, Department of Education, Office of Instruction and Program Development, and State of New York, Division of Criminal Justice Services.

Steffensmeier, D. (1986). *The fence: In the shadow of two worlds.* Totowa, NJ: Rowman and Littlefield.

Stockwell, T. (1997). Regulation of the licensed drinking environment: A major opportunity for crime prevention. In R. Homel (Ed.), *Policing for prevention: Reducing crime, public intoxication and injury* (pp. 7–33). Monsey, NY: Criminal Justice Press.

Sullivan, R. R. (2000). *Liberalism and crime: The British experience.* Lanham, MD: Lexington.

Sutherland, E. H. (1933). *The professional thief, by a professional thief.* Chicago: University of Chicago Press.

Sutherland, E. H. (1939). *Principles of criminology.* Chicago: Lippincott.

Sutton, M. (1995). Supply by theft: Does the market for second-hand goods play a role in keeping crime figures high? *British Journal of Criminology, 35,* 400–416.

Sutton, M. (1998). *Handling stolen goods and theft: A market reduction approach* (Home Office Research Study No. 178). London: British Home Office Research Publications.

Sutton, M. (2004). How burglars and shoplifters sell stolen goods in Derby: Describing and understanding the local illicit markets. *Internet Journal of Criminology,* pp. 1–44.

Sutton, M., Schneider, J., & Hetherington, S. (2001). *Tackling theft with the market reduction approach.* London: Policing and Reducing Crime Unit, U.K. Home Office.

Tatum, D. (Ed.). (1993). *Understanding and managing bullying.* London: Heineman.

Taylor, R. (2000). *Breaking away from broken windows.* Boulder, CO: Westview.

Tedeschi, J., & Felson, R. B. (1994). *Violence, aggression and coercive action.* Washington, DC: American Psychological Association.

Tittle, C. R., & Paternoster, R. (2000). *Social deviance and crime: An organizational and theoretical approach.* Los Angeles: Roxbury.

Toby, J. (1995). The schools. In J. Q. Wilson & J. Petersilia (Eds.), *Crime* (pp. 141–170). San Francisco: ICS Press.

Townsley, M., Homel, R., & Chaseling, J. (2003). Infectious burglaries: A test of the near repeat hypothesis. *British Journal of Criminology, 43,* 615–633.

Tremblay, P. (1986). Designing crime. *British Journal of Criminology, 26,* 234–253.

Tremblay, P. (1993). Searching for suitable co-offenders. In R. V. Clarke & M. Felson (Eds.), *Routine activity and rational choice: Advances in criminological theory* (Vol. 5, pp. 17–35). New Brunswick, NJ: Transaction Books.

Tremblay, P., Clermont, Y., & Cusson, M. (1994). Jockeys and joyriders: Changing patterns in car theft opportunity structures. *British Journal of Criminology, 34,* 307–321.

Tremblay, P., & Morselli, C. (2000). Patterns in criminal achievement: Wilson and Abrahamse revisited. *Criminology, 38*(2), 633–659.

Tremblay, P., & Pare, P. (2003). Crime and destiny: Patterns in serious offenders' mortality rates. *Canadian Journal of Criminology and Criminal Justice, 45*(3), 299–326.

Vancouver Planning Department. (1999). *Vancouver urban design: A decade of achievements.* Vancouver, BC: Urban Design and Development Center.

van Dijk, J. J. (1994). Understanding crime rates: On interactions between rational choices of victims and offenders. *British Journal of Criminology, 34*(2), 105–121.

van Dijk, J. J. M., van Kesteren, J. N., & Smit, P. (2008). *Criminal victimization in international perspective: Key findings from the 2004–2005 ICVS and EU ICS.* The Hague, Netherlands: Boom Legal Publishers.

van Gemert, F., & Verbraeck, H. (1994). Snacks, sex, and smack: The ecology of the drug trade in the inner city of Amsterdam. In E. Leuw & I. H. Marshall (Eds.), *Between prohibition and legalization: The Dutch experiment in drug policy* (pp. 145–167). Amsterdam: Kugler.

von Hirsch, A. (1987). *Past or future crimes: Deservedness and dangerousness in the sentencing of criminals.* New Brunswick, NJ: Rutgers University Press.

von Hirsch, A., Garland, D., & Wakefield, A. (Eds.). (2000). *Ethical and social issues in situational crime prevention.* Oxford, UK: Hart.

Walsh, D. P. (1994). The obsolescence of crime forms. In R. V. Clarke (Ed.), *Crime prevention studies* (Vol. 2, pp. 149–163). Monsey, NY: Criminal Justice Press.

Warr, M. (1988). Rape, burglary and opportunity. *Journal of Quantitative Criminology, 4*(3), 275–288.

Warr, M. (2001). Age, peers and delinquency. In J. G. Weis, R. D. Crutchfield, & G. S. Bridges (Eds.), *Juvenile delinquency readings* (pp. 135–140). Thousand Oaks, CA: Pine Forge.

Warr, M. (2002). *Companions in crime: The social aspects of criminal conduct.* New York: Cambridge University Press.

Webb, B. (1997). Steering column locks and motor vehicle theft: Evaluations from three countries. In R. V. Clarke (Ed.), *Situational crime prevention: Successful case studies* (2nd ed., pp. 46–58). New York: Harrow & Heston.

Webb, B. (2007, February 16). Obituary, Barry Poyner. *The Independent.* Available: http://www.independent.co.uk/news/obituaries/barry-poyner-436547.html

Webster, S. D. (2005). Pathways to sexual offense recidivism following treatment: An examination of the Ward and Hudson self-regulation model of relapse. *Journal of Interpersonal Violence, 20*(10), 1175–1196.

Weisburd, D. (2005). Hot spots policing experiments and criminal justice research: Lessons from the field. *Annals of the American Academy of Political and Social Science, 599,* 220–245.

Weisburd, D., Bushway, S., Lum, C., & Yang, S. (2004). Trajectories of crime at places: A longitudinal study of street segments in the city of Seattle. *Criminology, 42*(2), 283–321.

Weisburd, D. L., & Eck, J. (2004). What can police do to reduce crime, disorder and fear? *Annals of the American Academy of Political and Social Science, 593,* 42–65.

Weisburd, D., & Green, L. (1995). Policing drug hot spots: Findings from the Jersey City DMA experiment. *Justice Quarterly, 12*(4), 711–735.

Weisburd, D., Waring, E., & Chayet, E. (2001). *White-collar crime and criminal careers.* Cambridge, UK: Cambridge University Press.

Weisburd, D., Wheeler, S., Waring, E., & Bode, N. (1991). *Crimes of the middle classes.* New Haven, CT: Yale University Press.

Weisburd, D., Wyckoff, L. A., & Ready, J. (2006). Does crime just move around the corner? A controlled study of spatial displacement and diffusion of crime control benefits. *Criminology, 44*(3), 549–591.

Weisel, D. L. (2002). *Burglary of single-family houses.* Washington, DC: U.S. Department of Justice, Office of Community Oriented Policing Services.

Weisel, D. L. (2005). *Analyzing repeat victimization.* Washington, DC: U.S. Department of Justice, Office of Community Oriented Policing Services.

West, D. J. (1993). *Male prostitution.* New York: Harrington Park Press.

Wheeler, S. (1983). White collar crime: History of an idea. In S. H. Kadish (Ed.), *Encyclopedia of crime and justice* (pp. 1652–1656). New York: Free Press.

White, R. C. (1932). The relation of felonies to environmental factors in Indianapolis. *Social Forces, 10,* 498–513.

Widom, C. S., & Maxfield, M. G. (2001). *An update on the cycle of violence, research in brief.* Washington, DC: U.S. Department of Justice, Office of Justice Programs, National Institute of Justice.

Wigginton, E. (1972). *The Foxfire book: Hog dressing, log cabin building, mountain crafts and foods, planting by the signs, snake lore, hunting tales, faith healing, moon.* Garden City, NY: Doubleday.

Wigginton, E. (1979). *Foxfire 5: Ironmaking, blacksmithing, flintlock rifles, bear hunting, and other affairs of plain living.* Garden City, NY: Doubleday.

Wigginton, E. (1999). *Foxfire 11: The old homeplace, wild plant uses, preserving and cooking food, hunting stories, fishing, and more affairs of plain living.* Garden City, NY: Doubleday.

Wikstrom, P.-O. (1985). *Everyday violence in contemporary Sweden: Situational and ecological aspects* (Report No. 2015). Stockholm: National Council for Crime Prevention.

Wikstrom, P.-O. (1995). Preventing city-center street crimes. In M. Tonry & D. Farrington (Eds.), *Crime and justice: A review of research* (Vol. 19). Chicago: University of Chicago Press.

Wilson, J. Q., & Abrahamse, A. (1992). Does crime pay? *Justice Quarterly, 9,* 359–377.

Wilson, J. Q., & Kelling, G. L. (1982, March). Broken windows: The police and neighborhood safety. *Atlantic Monthly,* pp. 2029–2038.

Winchester, S., & Jackson, H. (1982). *Residential burglary: The limits of prevention* (Research Study #74). London: British Home Office Research Publications.

Wortley, R. (1997). Reconsidering the role of opportunity in situational crime prevention. In G. Newman, R. V. Clarke, & S. G. Shoham (Eds.), *Rational choice and situational crime prevention* (pp. 65–81). Dartmouth, UK: Ashgate.

Wortley, R. (1998). A two-stage model of situational crime prevention. *Studies on Crime and Crime Prevention, 7*(2), 173–188.

Wortley, R., & Smallbone, S. (Eds.). (2006). *Situational prevention of child sexual abuse.* Monsey, NY: Criminal Justice Press.

Wright, J. P., & Cullen, F. P. (2000). Juvenile involvement in occupational delinquency. *Criminology, 38*(3), 863–896.

Wright, R. T., & Decker, S. H. (1997). *Armed robbers in action: Stickups and street culture.* Boston: Northeastern University Press.

Wrong, D. H. (1961). The oversocialized conception of man in modern sociology. *American Sociological Review, 26,* 183–193.

Zahm, D. (2004). Brighter is better. Or is it? The devil is in the details. *Criminology & Public Policy, 3*(3), 535–545.

Zahm, D., & Perrin, D. (1992). Safe for study: Designing the campus environment. *Journal of Security Administration, 15*(2), 77–90.

Zhang, S., & Chin, K. (2002). Enter the dragon: Inside Chinese human smuggling organizations. *Criminology, 40*(4), 737–768.

Zhang, S. X., Chin, K., & Miller, J. (2007). Women's participation in Chinese transnational human smuggling: A gendered market perspective. *Criminology, 45*(3), 699–733.

Zimring, F. F. (2001). *American youth violence.* New York: Oxford University Press.

Zimring, F. F., & Hawkins, G. (1999). *Crime is not the problem: Lethal violence in America.* New York: Oxford University Press.

Zolotor, A. J., & Runyan, D. K. (2006). Social capital, family violence and neglect. *Pediatrics, 117*(6), 1124–1131.

INDEX

ABOUT THE AUTHORS

Marcus Felson (PhD, University of Michigan) is Professor of Criminal Justice at Rutgers University. He is author of *Crime and Nature* (Sage Publications) as well as "A Theory of Co-Offending" (*Crime Prevention Studies*, Volume 15, 2003) and "Redesigning Hell: Preventing Crime and Disorder at the Port Authority Bus Terminal" (*Crime Prevention Studies*, Volume 6, 1997). With Ronald V. Clarke, he has co-authored *Opportunity Makes the Thief.* Professor Felson is the originator of the routine-activity approach to crime rate analysis and has been a guest lecturer in many nations, including Argentina, Australia, Belgium, Brazil, Canada, Chile, the Czech Republic, Denmark, England, Estonia, Finland, France, Germany, Hungary, Italy, Japan, Mexico, the Netherlands, New Zealand, Norway, Poland, Scotland, Spain, Sweden, and Turkey. He has given talks on crime to applied mathematicians at four universities, including UCLA and the Centro di Ricerca Matematica at the Ennio De Giorgi Scuola Normale Superiore in Pisa, Italy.

Rachel Boba (PhD, Arizona State University) is an Associate Professor at Florida Atlantic University. Her expertise includes crime prevention, crime analysis, problem-oriented policing, and police accountability. She is author of *Crime Analysis With Crime Mapping* (Sage Publications), one of the first books to provide specific techniques and examples for students and practitioners preparing to enter the crime analysis profession. In addition, her collaborative research with the Port St. Lucie (Florida) Police Department has earned two prestigious awards—the International Association of Chiefs of Police Excellence in Law Enforcement Research Award (2008) and Finalist for the Herman Goldstein Award for Excellence in Problem-Oriented Policing (2006).